Astrophysical Polarimetry

Far-infrared through millimeter wavelengths

Online at: https://doi.org/10.1088/2514-3433/ae1022

AMERICAN ASTRONOMICAL SOCIETY IOP | ebooks™

AAS Editor in Chief

Ethan Vishniac, Johns Hopkins University, Maryland, USA

About the program:

AAS-IOP Astronomy ebooks is the official book program of the American Astronomical Society (AAS) and aims to share in depth the most fascinating areas of astronomy, astrophysics, solar physics and planetary science. The program includes publications in the following topics:

- GALAXIES AND COSMOLOGY
- INTERSTELLAR MATTER AND THE LOCAL UNIVERSE
- STARS AND STELLAR PHYSICS
- EDUCATION, OUTREACH, AND HERITAGE
- HIGH-ENERGY PHENOMENA AND FUNDAMENTAL PHYSICS
- THE SUN AND THE HELIOSPHERE
- THE SOLAR SYSTEM, EXOPLANETS, AND ASTROBIOLOGY
- LABORATORY ASTROPHYSICS, INSTRUMENTATION, SOFTWARE, AND DATA

Books in the program range in level from short introductory texts on fast-moving areas, graduate and upper-level undergraduate textbooks, research monographs, and practical handbooks.

For a complete list of published and forthcoming titles, please visit iopscience.org/books/aas.

About the American Astronomical Society

The American Astronomical Society (aas.org), established 1899, is the major organization of professional astronomers in North America. The membership (~7,000) also includes physicists, mathematicians, geologists, engineers, and others whose research interests lie within the broad spectrum of subjects now comprising the contemporary astronomical sciences. The mission of the Society is to enhance and share humanity's scientific understanding of the universe.

Editorial Advisory Board

Astrophysical Polarimetry

Far-infrared through millimeter wavelengths

David T Chuss
Department of Physics, Villanova University, Villanova, PA, USA

Edward J Wollack
NASA Goddard Space Flight Center, Greenbelt, MD, USA

IOP Publishing, Bristol, UK

ISBN 978-0-7503-2753-4 (ebook)
ISBN 978-0-7503-2751-0 (print)
ISBN 978-0-7503-2754-1 (myPrint)
ISBN 978-0-7503-2752-7 (mobi)

DOI 10.1088/2514-3433/ae1022

Version: 20260301

AAS–IOP Astronomy
ISSN 2514-3433 (online)
ISSN 2515-141X (print)

British Library Cataloguing-in-Publication Data: A catalogue record for this book is available from the British Library.

Published by IOP Publishing, wholly owned by The Institute of Physics, London

IOP Publishing, No.2 The Distillery, Glassfields, Avon Street, Bristol, BS2 0GR, UK

US Office: IOP Publishing, Inc., 190 North Independence Mall West, Suite 601, Philadelphia, PA 19106, USA

To our collaborators and colleagues who share the wonder of nature and the passion for exploration.

Contents

Acknowledgements

I am deeply grateful to my wife Allison and our sons, whose love, patience, and constant encouragement have carried me through every stage of this work. Their belief in me—and their willingness to support the many hours spent thinking, writing, and wandering through ideas—made this book possible. My sincere thanks go to my coauthor Dave, whose insight, dedication, and steady partnership have enriched every chapter. It has been a privilege to explore new intellectual terrain together and to collaborate so closely on this shared vision. Over the years, many colleagues have shaped my thinking through generous conversations, debates, and shared curiosity. These discussions—sometimes brief, sometimes sprawling—pushed me to see problems differently and to continue the long, rewarding journey of learning. I would also like to acknowledge our editors, whose thoughtful guidance, observations, and careful attention strengthened this work in countless ways. Much of this project was built in moments of quiet exploration, sitting at a computer and venturing outward through ideas, data, and the accumulated knowledge of others. In those moments, I often thought of Arthur C. Clarke's reminder that "...the only thing we can be sure of about the future is that it will be different...." That spirit of curiosity and openness—of exploring what is possible and what possibilities lie ahead—has guided both this book and my broader work. To all a sincere thank you. This book exists because of your support and contributions. — EJW / 2025.11.16

I would like to express my appreciation for the love and support that Anne Marie and my children, Katie, Tommy, and Mia have offered through the writing of this manuscript. Without their belief in me, I could not have completed this project. I would like to thank Tommy specifically, for a careful read and subsequent suggestions on early drafts of several of the chapters. I would like to thank my parents, Tom and Nancy for their constant belief in me and unwavering support. Reflecting my co-author's comments, I would like to thank Ed in his roles as a mentor, colleague and friend. This book reflects so many of our conversations over the years that have enabled me to see things differently and make connections that have made me a better scientist. Everyone's journey is due to so much beyond their own talents and effort and cannot be successfully undertaken without good mentorship and collaboration along the way. I have been fortunate to have both. I'm forever grateful for my PhD advisor, Giles Novak and for my postdoctoral advisor Harvey Moseley, my colleagues at NASA Goddard and at Villanova along with the myriad of collaborators with whom I've enjoyed working on a variety of projects. In addition, I would like to thank my many students and postdocs along the way from whom I've learned more than I could have possibly taught them. — DTC / 2025.11.29.

Author biographies

David T. Chuss

David Chuss received his BS in Physics from Villanova University in 1995 and then went on to earn an MS in Astronomy and Astrophysics from Penn State. He followed this with a PhD in Physics and Astronomy from Northwestern University. From there, he moved to NASA Goddard Space Flight Center, first as a National Research Council Postdoctoral Associate, and then as a Research Astrophysicist. In January 2015, he returned to Villanova University, where he is now a Professor and Chair of the Department of Physics.

Dr. Chuss has continuously worked in the field of astronomical polarimetry throughout his career. As such, he has developed instrumentation to study the polarization of long wavelength light from astrophysical phenomenon to learn about the universe. He has used this technique to study the physics of magnetic fields in our Milky Way's Galactic center and has applied these techniques to study the origins of the universe. His work has been publicized in the popular press in places such as *The New York Times*, *Scientific American*, Space.com, and *Forbes*. His instrumentation efforts have led to several adventures including five trips to the South Pole and numerous trips to mountain tops in Hawaii, Arizona, and Chile.

Dr. Chuss has been a team member for several projects including the HAWC+ polarimeter for NASA's Stratospheric Observatory for Infrared Astronomy (SOFIA), an airborne telescope, and the Cosmology Large Angular Scale Surveyor (CLASS), a polarimeter constructed in the Atacama Desert in Chile and led by Johns Hopkins University to test the theory of cosmic inflation.

Edward J. Wollack

Ed Wollack is a research astrophysicist at the NASA Goddard Space Flight Center (GSFC). He received a Bachelor of Science degree in physics from the University of Minnesota (1987). After completing his doctoral degree in physics at Princeton University (1994) he joined the research staff at the National Radio Astronomy Observatory's Central Development Laboratory where he concentrated on the development of millimeter wave receivers for precision astrophysical observations. In 1998, he joined the GSFC to implement the Wilkinson Microwave Anisotropy Probe (WMAP) mission. In 2014, Dr. Wollack joined the Wide-Field Infrared Survey Telescope project as a study scientist and was named the Roman Space Telescope Observatory project scientist (2018). Within the NASA Engineering and Safety Center he has contributed to resolution of sensor and instrumentation challenges encountered in development and application of complex systems.

Dr. Wollack has contributed broadly to the scientific and technical literature on topics including cosmology, astrophysics, and instrumentation. He has designed and implemented sensors, components, and instrument systems for a wide variety of ground, sub-orbital, and space-borne applications. These research contributions have spanned the electromagnetic spectrum. His professional recognitions include NASA's Exceptional Service Award, Exceptional Achievement Medal, and induction into the Inventor Hall of Fame. He is a recipient of the Gruber Prize in Cosmology, the Breakthrough Prize for Fundamental Physics, and the National Science Foundation Antarctic Service Medal.

Dr. Wollack is an avid contributor to education and public outreach in venues ranging from local classrooms to permanent museums. He has given numerous invited public presentations, developed hands-on science and math activities, and served as an editor and content advisor for NASA's award winning "Tour of the Electromagnetic Spectrum" animated series. He has worked with curators at institutions including the Deutsches Museum München, the Smithsonian National Air and Space Museum, and the Goddard Visitor Center to develop science content for public exhibit. Ed enjoys solving puzzles and is fascinated by light, its interaction with matter, and its role in revealing patterns in our Universe.

Chapter 1

Introduction

Why polarimetry? Astrophysics involves the observation, characterization, and interpretation of light emitted over cosmic time. These photons—whether emitted by a planet in our vicinity or the relic plasma from the Big Bang at immense distance —encode a story of the conditions of their origin and their interaction with the contents of the Universe. We utilize these measurements to infer details about the physical processes that produced or otherwise affected that light along its journey to us. In interpreting these signatures it is not uncommon to implicitly assume that the underlying physical properties of light and matter follow the rules observed in our humble terrestrial laboratory settings. As a result, the Universe's light extends the distance, time, and energy scales over which its contents can ultimately be sensed and explored. This is a form of remote sensing that provides a testable means for extending our theoretical understanding of the fundamental forces of nature and their influence over space and time.

Imaging and spectroscopy are measurement techniques that characterize the intensity of the light as a function of its direction and energy, respectively. Historically, such information has proven invaluable in revealing and resolving the mysteries of the cosmos. Imaging and spectroscopy alone do not consider the vector nature of light, which is ultimately traceable to the angular momentum of the (spin 1) photon. Hence these techniques do not utilize all of the information that could be brought to bear in the effort to increase our understanding of the Universe.

The concept of *polarization* is a manifestation of the vector nature of light, and its measurement provides additional information about the physical processes involved in a photon's history. It is tempting to think of polarimetry as a separate technique from spectroscopy and imaging; however, a more complete treatment considers polarimetric quantities to be treated in concert with both imaging and spectroscopy. In other words, the vector properties of light can be measured as a function of frequency and direction in space.

doi:10.1088/2514-3433/ae1022ch1

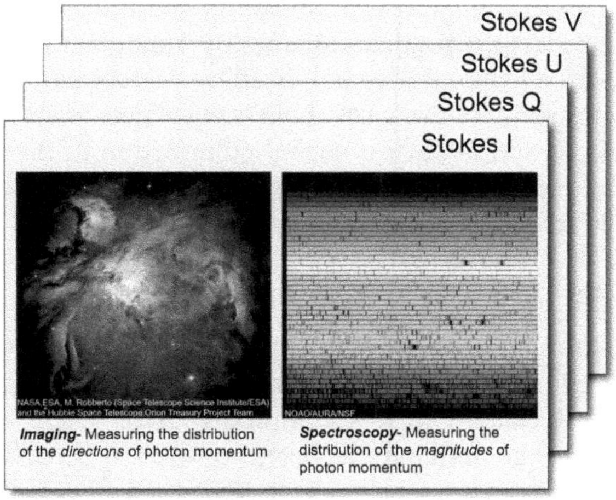

Figure 1.1. Astrophysical measurements can be broadly divided into two categories: Imaging, in which the distribution of the directions of the photon momenta are measured and spectroscopy, in which the distribution of the magnitudes of photon momenta are measured. More generally, to represent the vector nature of light, four distinct parameters need to be specified. These quantities provide additional information about the physical history of the photons and their interaction with matter.

We will see in the next chapter that the polarization requires four independent parameters to fully describe the nature of light at a single point and frequency (see Figure 1.1). The total intensity, a scalar quantity, is just one of these four parameters. Maps and spectra can be measured for all four of these so-called "Stokes parameters," so from this perspective, without polarimetry only a subset of the information space provided by incoming radiation in astrophysics can be explored. If we could map the four Stokes parameters over the entire sky at infinite angular and spectral resolution, we would have all of the information we can obtain.[1]

Light, being an electromagnetic wave, consists of electric and magnetic fields oriented in the plane perpendicular to the direction of propagation in free space. For monochromatic light on infinitesimal timescales, light will always be 100% polarized, since one can, in principle, measure the instantaneous magnitude and direction of a classical field. However, real sources of light (and the detectors we use) have finite spectral bandwidths, temporal response, and physical variability. In addition, the phases of the emitting radiation can be random in time and spatially correlated. This leads to the consideration of the statistical nature of polarization. When a physical process in space correlates the phases of the polarization components of the relevant radiation field, the light, when averaged over an appropriate time interval, can exhibit *partial polarization*. A measurement sensitive to polarization can in turn convey information about the phenomenon in question.

Magnetic fields are an example of such a physical phenomenon. Interstellar dust grains can become aligned in the presence of these fields, and in turn both their

[1] Note that this statement applies to a given instant in time, as the Universe is dynamic.

extinction and emission can become partially-polarized. Synchrotron emission, radiation emitted by charged particles aligned by a magnetic field and accelerated perpendicular to their velocity, is also polarized by its very nature. The geometry of scattering off of electrons in the early Universe, which is set by the primordial distribution of dark matter, imparts a partial polarization to the cosmic microwave background. In each of these cases, measuring the polarization provides information that is not available from the intensity of light or its spectrum.

1.1 A Tour of the Long Wavelength Spectrum

Representative spectra for the total intensity of continuum sources that are known to be partially-polarized are shown in Figure 1.2. This figure also shows rough boundaries of the wavelength ranges commonly used. There are no universally accepted wavelength range designations here; however, it is useful to have a rough idea of these different regimes as they tend to lightly correlate with the physics they probe, and their relevant instrumentation techniques and technologies are often shared with those of their spectral neighbors. These device and system implementations are largely driven by readily available materials, manufacturing capability, and experimental ingenuity. As a result, these strategies merely represent a snapshot in time and looking toward the future are certain to continue to evolve.

1.2 A Brief History: Polarimetry Highlights

The history of polarimetry has been extensively chronicled by Brosseau (1998). Here we provide a brief summary of polarimetry highlights as it relates to the development of long wavelength astrophysical observations.

- 1621 Snell: Observation of total internal reflection in transparent materials.
- 1669 Bartholinus: Discovery of double refraction in calcite.

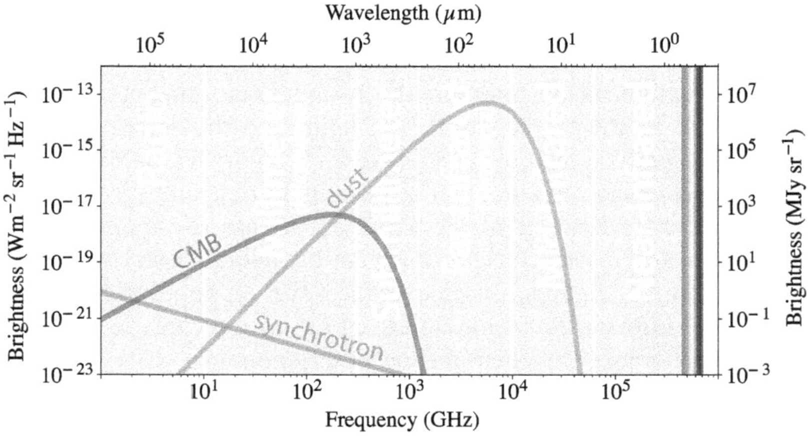

Figure 1.2. Rough boundaries of the different wavelength/frequency regimes in long wavelength astrophysics are shown. Three main types of continuum sources that exhibit polarization are shown using representative spectra: synchrotron emission, thermal emission from dust, and the cosmic microwave background.

- 1670 Huygens: Interpretation of calcite birefringence.
- 1672 Newton: Coins term "polarization" due to "attractive virtue lodged in certain sides" by analogy with magnetic poles.
- 1808 Malus: Discovery that light becomes polarized upon reflection.
- 1809 Arago: First reported observation of the polarization of the skylight.
- 1810 Malus: Discovery of transmitted intensity extinction by crossed polarizer and Malus' cosine squared law.
- 1815 Brewster: Discovery of plane-polarized light reflection angle.
- 1823 Fresnel: Derivation of formulas for reflection and refraction in electromagnetic media.
- 1828 Nicol: Invention of the Nicol prism, which enabled microscope optics with a single-polarization. The concept of extinction via a crossed polarizers represents a natural extension for analyzing samples with such instruments.
- 1840 Faraday: Discovery of polarization rotation in optically active media.
- 1851 Stokes: Introduction of Q, U, V and I description of partial polarization.
- 1864 Maxwell: Unified electromagnetic field concept (Maxwell 1865).
- 1897 Bose: demonstration of millimeter wave cut-off metal plate polarizer (i.e., sheets of foil interleaved between pages of "Bradshaw's Railway Timetable").
- 1923 Lyot: Performs polarimetry of sunlight scattered by Venus.
- 1942 Institute of Radio Engineers (IRE): Definition of standardized notation and sign convention for polarization.
- 1946 Chandrasekhar: Predicted electron-scattered starlight linear polarization.
- 1949 Hiltner and Hall: First measurement of starlight polarization.and discovery of interstellar polarization
- 1951 Land/Polaroid Corporation: Development of sheet polarizers.
- 1951 Davis and Greenstein: Realize dust magnetically aligned.
- 1953 Shklovskii: Proposes synchrotron as source of Crab's polarized radio emission and optical continuum (detection 1957).
- 1971 Clarke and Grainger: Discovery of dichroism (K_2PdCl_4).
- 1972 Columbia University: Polarized X-ray emission from Crab Nebula detected.
- 1973 International Astronomical Union (IAU): Endorses IEEE elliptical polarization definition and definition of polarization angle.
- 1976 Dolginov and Mitrofanov: Propose radiative alignment torque (RAT) theory.
- 1982 Cudlip et al: First detection of astronomical polarization in FIR emission.
- 2002 Degree Angular Scale Interferometer (DASI): Detection of Cosmic Microwave Background (CMB) polarization.

1.3 The Organization of This Book

In this book, we primarily focus on the far-infrared through millimeter portion of the electromagnetic spectrum. These bands are also referred to as the "terahertz" and "microwave," respectively. From a measurement standpoint, we will focus on instrumentation for single-dish telescopes. For coherent receivers and imaging systems, a larger range of polarimetric implementation and calibration options exist. For details regarding interferometric arrays, see Thompson et al. (2017) and references therein. Where possible, we will provide breadcrumbs and references for the interested reader to explore beyond the topics we cover.

Chapter 2 presents formal concepts necessary for polarimetry, introducing the definition of polarization as a manifestation of the partial coherence of natural light. We introduce the Stokes parameters and the Poincaré Sphere as a graphical approach to visualizing them. In Chapter 3, we discuss sources of polarization, and focus on matter-radiation interactions relevant to both the physical processes we wish to measure and the instrumentation we use to measure them. In Chapter 4, we discuss the measurement of polarization from both system and device perspectives. In Chapter 5, we discuss architectures of polarimeters. We include a brief history of polarimeters in the far-infrared and submillimeter. Chapter 6 focuses on tools for dealing with data, in both reduction and scientific analysis. Finally, we discuss science topics and discoveries in Chapter 7.

References

Brosseau, C. 1998, Fundamentals of Polarized (1st ed.; New York: Wiley)

Maxwell, J. C. 1865, RSPT, 155, 459

Thompson, A. R., Moran, J. M., & Swenson, G. W. 2017, Interferometry and Synthesis in Radio Astronomy (Berlin: Springer)

Chapter 2

Polarization Theory

Polarization of light is a natural consequence of the vector nature of the electromagnetic field. Monochromatic plane wave solutions to Maxwell's wave equation are by their nature 100% polarized since each mode is (self-)coherent, and the vector sum of polarized waves leads to a new vector solution. In practice, light sources arising from ensembles of charges in motion are not generally coherent. *Incoherent* light is a result of the superposition of waves having random phases over a finite bandwidth. In general though, the phase may not be completely random. Physical processes or interactions can partially correlate the fields, resulting in partial polarization. For example, radiation incident on a plane of closely-spaced metal wires will experience an induced phase correlation such that the reflected wave is polarized with its electric field parallel to the wires, and the transmitted wave is polarized perpendicular to the wires. Thus, a wire grid can be used as a polarization filter. Alternatively, plane-polarized light will diffusely scatter from a roughened surface into a spectrum of plane-waves and thus become decorrelated. Lambertian surfaces and blackbody cavities provide idealized examples of this behavior, which lead to unpolarized light. Temporal and spatial variations in the phase provide paths to influence the degree of polarization.

The polarization state of light is most generally a *stochastic* physical phenomenon. It is applicable for signals that are *ergodic*, that is, those for which the time average and ensemble averages are equivalent. Another way of stating this is that the statistical properties of the radiation field are essentially constant for a time interval that is larger than the inverse of the bandwidth of the radiation (the correlation time). For a more complete treatments of ergodic analytic signals see Brosseau (1998) and Wolf (2007).

2.1 Electromagnetic Waves in Free Space

We start this section with a brief review of the electromagnetic wave solutions of Maxwell's equations in a source-free medium. We discuss polarization as a

doi:10.1088/2514-3433/ae1022ch2

description of the vector nature of electromagnetic radiation. We will then be able to expand this picture to include partially-coherent radiation fields, at which point we generalize our description of polarization.

The source-free Maxwell's equations can be expressed using MKS units as follows:

$$\begin{aligned} \nabla \cdot \mathbf{E} = 0 \quad & \nabla \times \mathbf{E} + \mu_o \frac{\partial \mathbf{H}}{\partial t} = \mathbf{0} \\ \nabla \cdot \mathbf{H} = 0 \quad & \nabla \times \mathbf{H} - \varepsilon_o \frac{\partial \mathbf{E}}{\partial t} = \mathbf{0}. \end{aligned} \tag{2.1}$$

where ε_o and μ_o are the permittivity and permeability of freespace, respectively. By convention, polarization of the electromagnetic field is associated with the orientation of the electric field vector (or more generally, the correlations of its components). This is convenient because most interactions can be considered *electric* in nature. This is perhaps related to the fact that while electric charges exist, magnetic monopoles do not (as far as is known from observation). From a relativistic perspective, the magnetic field can be understood as a transformation of the electric field (or vice versa), dependent upon the observer's frame of reference. At a fundamental level, classical electromagnetic fields are intimately linked to charge separation and its variation as a function of time.

In such source-free regions, it can be shown that Maxwell's equations imply that the electric field is the solution of the wave equation in three dimensions,

$$\nabla^2 \mathbf{E} - \varepsilon_o \mu_o \frac{\partial^2 \mathbf{E}}{\partial t^2} = \frac{\partial^2 \mathbf{E}}{\partial x^2}\hat{x} + \frac{\partial^2 \mathbf{E}}{\partial y^2}\hat{y} + \frac{\partial^2 \mathbf{E}}{\partial z^2}\hat{z} - \varepsilon_o \mu_o \frac{\partial^2 \mathbf{E}}{\partial t^2} = \mathbf{0}, \tag{2.2}$$

where $c = \sqrt{1/\varepsilon_o \mu_o}$ is the speed of light in free space. An equivalent wave equation exists for the magnetic field. The electric field vector of the electromagnetic wave, which defines its polarization, is given by the solution to the above equation,

$$\mathbf{E}(\mathbf{r}, t) = \hat{\mathbf{e}}\mathcal{E}(\mathbf{r})e^{i\mathbf{k}\cdot\mathbf{r} - \omega t}, \tag{2.3}$$

where $\mathcal{E}(\mathbf{r})$ is a complex-valued function of position that incorporates the phase of the electromagnetic wave. Though this formalism is general for any mode set, as an example and $\hat{\mathbf{e}}$ is a complex unit vector, we can write the expression for plane waves traveling in the $+z$-direction,

$$\mathbf{E}(\mathbf{r}, t) = \left(E_x\hat{\mathbf{x}} + E_y\hat{\mathbf{y}}\right)e^{ikz - i\omega t}, \tag{2.4}$$

where E_x and E_y are complex amplitudes that allow for a phase difference between the two polarization components. We have chosen to express the field in Cartesian coordinates; however, any basis in the plane perpendicular to the direction of propagation can be utilized. This includes any bases related to the $(\hat{\mathbf{x}}, \hat{\mathbf{y}})$ basis by a complex linear combination. As an example, we can express the independent fields in terms of the circular polarization basis,

$$\mathbf{E}(\mathbf{r}, t) = (E_r\hat{\mathbf{e}}_r + E_\ell\hat{\mathbf{e}}_\ell)e^{ikz - i\omega t}, \tag{2.5}$$

Here, $\hat{\mathbf{e}}_r = \frac{(\hat{\mathbf{x}} + i\hat{\mathbf{y}})}{\sqrt{2}}$ and $\hat{\mathbf{e}}_\ell = \frac{(\hat{\mathbf{x}} - i\hat{\mathbf{y}})}{\sqrt{2}}$, are the right and left circular polarization basis vectors, respectively.

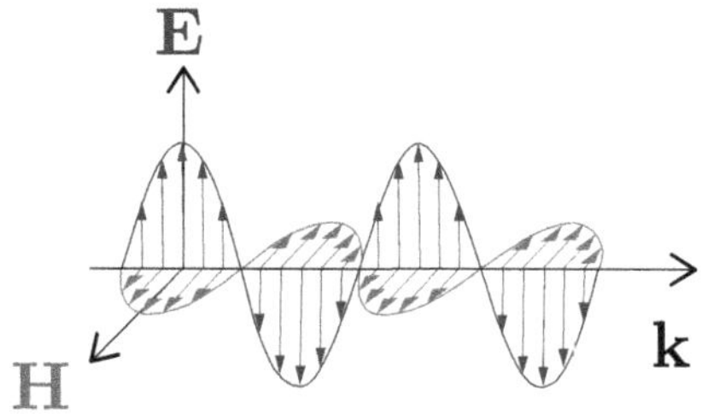

Figure 2.1. In free space, the **E**, **H**, and **k** vectors form an orthogonal set. Here, the relationship is shown for a linearly-polarized plane wave.

From the laws of Faraday and Ampere, we know that the relation between the electric and magnetic fields and the direction of propagation, $\hat{\mathbf{k}}$, is given by

$$\mathbf{E} = \sqrt{\frac{\mu}{\varepsilon}}\hat{\mathbf{k}} \times \mathbf{H} = \eta\hat{\mathbf{k}} \times \mathbf{H}, \tag{2.6}$$

where η is the wave impedance of the medium in which the wave is propagating. Thus, the electric and magnetic fields exist in the plane perpendicular to the direction of propagation. An example of this relationship between these vectors is shown in Figure 2.1.

Use of Magnetic Field (H) in Radiation Theory

It is common to utilize the magnetic field **H** rather than the magnetic induction, **B**, in treating radiating fields. While either can be accommodated, a useful consequence of this choice is the identification of [**E**, **H**] with voltage and current ([V, I]) from the integral forms of the electromagnetic field equations:

$$I = \oint \mathbf{H} \cdot d\boldsymbol{\ell}, \tag{2.7}$$

$$V = \int \mathbf{E} \cdot d\boldsymbol{\ell}. \tag{2.8}$$

For plane wave illumination, these transverse field amplitudes readily provide an additional insight: Defining their ratio as $\eta \equiv |E|/|H|$ and noting it has units of Ohms (Ω) serves to highlight the role of the medium's *wave impedance* in defining where electric (or magnetic) fields dominate the response and coupling to propagating waves.

Though both the **E** and **H** fields are present in electromagnetic radiation, it is customary to restrict a discussion of polarization to the direction of **E**. Because the electric and magnetic fields are connected via Equation (2.6), once the direction of one of the fields is determined, so is the other. Thus, keeping track of one of these fields is sufficient for understanding the propagation. The polarization is associated with the electric field by definition, a convention motivated by the observation that the interaction between electromagnetic radiation and matter is most commonly mitigated by the electric part of the radiation field. This was illustrated by a famous experiment of Wiener (1890), in which he found that photographic plates were exposed in regions of maximum electric fields and not exposed in the corresponding maxima of the magnetic field. Ultimately, this is a consequence of the observed asymmetry in Maxwell's equations between the existence of electric and magnetic charges and currents.

In much of what we do in the formalisms in this book, we will assume the *paraxial approximation*, which states that the rays going through an optical system are concentrated around a small range of angles around the central optical axis. The equivalent statement here (where we are concerned about phases, and thus the wave picture of light) is that the electric and magnetic fields are transverse to the central axis. We will see that from a polarization perspective, we can transform the fields in the plane perpendicular to the direction of propagation. As such, transformations can be described as complex linear combinations of the transverse field components or equivalently as an angular "spectrum of plane-wave" radiation. Such an approach motivates the use of linear algebra, matrices, and Fourier transforms. We will explore some of these techniques in the sections and chapters that follow.

In cases involving violations of this approximation, more sophisticated analysis methods may be brought to bear (Balanis 2024). For example, in designing radiating structures and guided wave components, numerical modeling methods which include the finite element method (FEM) (Jin 2015), finite difference time domain (FDTD) (Yee 1966), and method of moments (MoM) (Harrington 1987, 1993) are commonly employed. In optical systems, the polarization response can be computed by keeping track of the vector nature of the radiation via full-wave, diffractive, or geometric ray-tracing analysis. However, the paraxial approximation provides an insightful tool in developing an understanding of polarization at the component and system levels in modeling instrumentation for astronomical polarimetry.

2.2 The Polarization Ellipse

In visualizing polarization of monochromatic radiation, it is useful to consider the time-dependent resultant electric field in a plane that is perpendicular to the propagation direction. This concept appears in many places (Brosseau 1998; Goldstein 2003; Hecht 2017), and we will review this strategy here. As above, we choose a coordinate system such that the propagation vector is $\mathbf{k} = k\hat{\mathbf{z}}$ and consider the nature of the oscillations of the electric field in the x–y plane. For such a formalism, we require a total of three degrees of freedom: the amplitude of each orthogonal component and the relative phase between them. As before, we can pick any two orthogonal polarizations; however, it is intuitive to choose a linear polarization basis. In this case, we choose the x-y basis such that the amplitudes are E_{x0}, and E_{y0}, relative phase is defined as δ, and express the electromagnetic fields components as:

$$\begin{aligned} E_x &= E_{x0}\cos(kz - \omega t), \\ E_y &= E_{y0}\cos(kz - \omega t + \delta). \end{aligned}$$

One can expand the second expression using the fact that $\cos(\alpha \pm \beta) = \cos\alpha\cos\beta \mp \sin\alpha\sin\beta$,

$$\frac{E_y}{E_{y0}} = \cos(kz - \omega t)\cos\delta - \sin(kz - \omega t)\sin\delta. \tag{2.9}$$

From the first expression, one can see that

$$\cos(kz - \omega t) = \frac{E_x}{E_{x0}},$$
$$\sin(kz - \omega t) = \left(1 - \frac{E_x^2}{E_{x0}^2}\right)^{\frac{1}{2}}.$$

Putting these expressions into Equation (2.9), it is possible to remove the time dependence and express the path of the electric field in a plane perpendicular to the propagation direction,

$$\left(\frac{E_y}{E_{y0}}\right)^2 + \left(\frac{E_x}{E_{x0}}\right)^2 - 2\frac{E_x E_y}{E_{x0}E_{y0}}\cos\delta = \sin^2\delta. \tag{2.10}$$

Equation (2.10) describes the red elliptical path in Figure 2.2. The orientation of the semimajor axis of this ellipse is denoted by ϕ, which is chosen to conform to IAU[1] convention, measured counterclockwise from the north direction in a celestial coordinate system,

$$\tan 2\phi = \frac{2E_{x0}E_{y0}\cos\delta}{E_{x0}^2 - E_{y0}^2}. \tag{2.11}$$

As can be seen from this equation, ϕ is dependent on all three parameters of the plane wave, E_{y0}, E_{y0}, and δ.

IAU Polarization Convention

The polarization angle ϕ, via IAU convention (XVth General Assembly, Sydney, Australia, 1973), is measured counterclockwise from North in the relevant coordinate frame (be it Galactic, Ecliptic, etc.). However, care must be taken in that not all literature conforms to this convention. The HEALpix convention (Górski et al. 2005) is sometimes used (especially with studies of the cosmic microwave background radiation). In this standard, ϕ is measured *clockwise* from North. The difference between these two systems results in a sign flip in the Stokes U parameter.

[1] International Astronomical Union.

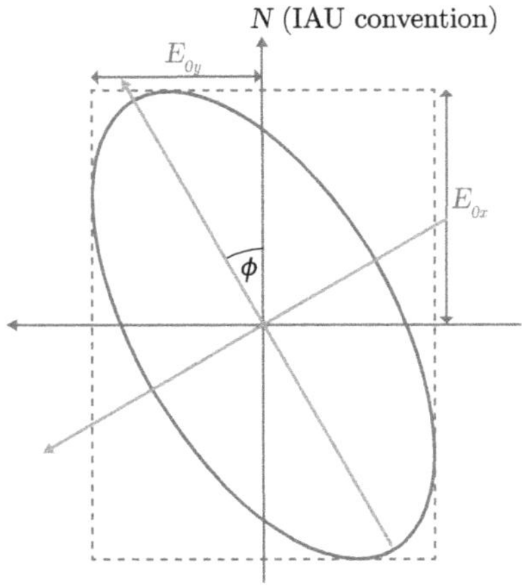

Figure 2.2. In the most general case, the electric field traces an ellipse in a plane perpendicular to the propagation direction. We follow the IAU convention that the semimajor axis of the ellipse is referenced counterclockwise from celestial north by the position angle, ϕ. We show the (green) coordinate system in which the amplitudes of two linear orthogonal components are equal. In this basis, the eccentricity of the ellipse is only dependent on the relative phase between the components.

2.3 Jones Vectors

The transverse fields at a single plane along the propagation direction can be represented by a two-component vector known as a *Jones vector* (Jones 1941). To represent this, we borrow the "bra-ket" (or Dirac) notation from quantum mechanics,

$$|E\rangle = \begin{pmatrix} E_x \\ E_y \end{pmatrix}.$$

The common $e^{i(kz-\omega t)}$ factor has been omitted here for simplicity. This compact notation and formalism has utility in situations where the light is fully coherent. Connecting this to the polarization ellipse, one finds that

$$|E\rangle = \begin{pmatrix} E_{x0} \\ E_{y0}\exp(i\delta) \end{pmatrix}.$$

The polarization eigenstates for monochromatic light can be represented by (normalized) Jones vectors as follows:

$$\begin{aligned}
|E\rangle &= \begin{pmatrix} 1 \\ 0 \end{pmatrix} \rightarrow \text{Horizontal Polarization} \\
|E\rangle &= \begin{pmatrix} 0 \\ 1 \end{pmatrix} \rightarrow \text{Vertical Polarization} \\
|E\rangle &= \frac{1}{\sqrt{2}}\begin{pmatrix} 1 \\ 1 \end{pmatrix} \rightarrow \text{Diagonal Polarization} \\
|E\rangle &= \frac{1}{\sqrt{2}}\begin{pmatrix} 1 \\ i \end{pmatrix} \rightarrow \text{Circular Polarization.}
\end{aligned}$$

2.4 Polarization Transformation in Optical Systems: Jones Matrices

Whether intentionally or otherwise, components of optical systems alter the coherence state of the radiation as it propagates through a given system. We will see later in this work that in polarimeters, the action of modulators and analyzers are designed to alter the polarization state of the system prior to detection. In this section, we will explore common matrix techniques to compute and track the polarization response of an instrument.

The polarization state of fully coherent light can be modified through either the change of amplitude of one polarization component relative to the other, or introduction of a phase delay between two orthogonal polarization states. The former can be achieved in optical systems using either *biattenuant* materials, or by diplexing the incoming light using a polarizer or equivalent device (see Chapter 4). The latter can be accomplished using a path delay such as in an interferometer or by employing *birefringent* materials, those having differing phase velocities for the two orthogonal polarization states. Examples of devices that employ differential phase delays include Faraday rotators and wave plates.

A simple approach to the transformation of polarization is to represent the action of an optical system on the transverse fields mathematically. In this case, we will assume the z-direction to be the direction of propagation, and we suppress the harmonic ($e^{i\omega t}$) factor. The operation of an optical system is assumed to be linear in this formalism. The fields at any point in a system can be obtained by a linear combination of the fields at the preceding point in the system. This allows us to represent the action of the system by a 2×2 complex matrix called a *Jones Matrix*,

$$\mathbf{J} = \begin{pmatrix} J_{11} & J_{12} \\ J_{21} & J_{22} \end{pmatrix}.$$

The transformation of a coherent field distribution, represented by a Jones vector, by an optical system, represented by Jones matrix, is given by

$$\begin{pmatrix} E_x' \\ E_y' \end{pmatrix} = \begin{pmatrix} J_{11} & J_{12} \\ J_{21} & J_{22} \end{pmatrix} \begin{pmatrix} E_x \\ E_y \end{pmatrix}.$$

Throughout this book we will formulate the vector nature of light using the linear polarization basis. In doing so, we can construct any arbitrary polarization state from a complex linear combination of (say) horizontal (x) and vertical (y) polarization components. However, it is possible and equally valid to choose right- and left-circular polarization as our basis. That is, we can re-do all of the math in this section as complex linear combinations of the circular polarization states.

We can also express the electric field in Dirac notation, with the "state" of the new electric field denoted $|E'\rangle$,

$$|E'\rangle = \mathbf{J}|E\rangle.$$

An example of the use of Jones' matrices is a coordinate rotation,

$$\begin{pmatrix} E'_x \\ E'_y \end{pmatrix} = \begin{pmatrix} J_{11} & J_{12} \\ J_{21} & J_{22} \end{pmatrix}\begin{pmatrix} E_x \\ E_y \end{pmatrix} = \begin{pmatrix} \cos\theta & -\sin\theta \\ \sin\theta & \cos\theta \end{pmatrix}\begin{pmatrix} E_x \\ E_y \end{pmatrix} = \begin{pmatrix} E_x\cos\theta - E_y\sin\theta \\ E_x\sin\theta + E_y\cos\theta \end{pmatrix}.$$

Thus, the new (primed) field components can be expressed in terms of the old (unprimed) field components,

$$\begin{aligned} E'_x &= E_x\cos\theta - E_y\sin\theta \\ E'_y &= E_x\sin\theta + E_y\cos\theta. \end{aligned}$$

The magnitude of the Poynting vector ($\mathbf{S} = \mathbf{E} \times \mathbf{H}$) is proportional to the square of the electric fields. In this formalism, the power is the inner product the Hermitian conjugate of the Jones vector with the vector itself, where for notational convenience a factor of $\frac{1}{2}\eta$ has been absorbed into the operational definition of $\mathscr{P}$,

$$\mathscr{P} = \mathbf{E}^{\dagger}\mathbf{E} = \begin{pmatrix} E_x^* & E_y^* \end{pmatrix}\begin{pmatrix} E_x \\ E_y \end{pmatrix} = E_x^2 + E_y^2.$$

Equivalently, bra-ket notation can also be used to denote power,

$$\mathscr{P} = \langle E|E\rangle = \begin{pmatrix} E_x^* & E_y^* \end{pmatrix}\begin{pmatrix} E_x \\ E_y \end{pmatrix} = E_x^2 + E_y^2.$$

2.5 Natural Light

Under the idealization of monochromatic light, to which we have limited the discussion up to this point, polarization is always well-defined and complete because by definition the light is fully coherent. That is to say that the phase of the beam is always well-defined. Monochromatic light is not found in nature, though it provides a reasonable approximation for things like lasers and masers. In general, light consists of a set of frequencies spanning a finite spectral range and duration. The phases of these constituent waves are in most cases random both spatially and temporally (as a result of the physics of their sources). This applies to both orthogonal linear polarizations.

Thermal blackbody radiation, a common example of natural light, is to a good approximation an unpolarized radiation source. Holding a nearly ideal polarizer up to a source of such light, one will generally find that there is no orientation of the polarizer that completely blocks all of the radiation. Half of the light is reflected (or absorbed) and half is transmitted. Such unpolarized light is *not* able to represented by a Jones matrix. More generally the concept of *partial polarization* is required to fully describe the properties of electromagnetic radiation found in nature. One can think of partially polarized light as a superposition of a component that is polarized (i.e., aligned) and one is not (i.e., random). Contrast the previous example with holding two polarizers in series and varying their relative orientation. In this second case, if the polarizers are perfect, there is an orientation of the polarizers for which all of the light is blocked. This is because after the first polarizer, the transmitted light is 100% polarized, such that the electric field is uniformly oriented. Thus the second polarizer can be oriented at 90° relative to the first to eliminate transmission. Physical processes that produce partially-polarized light are more common. For example, one could either scatter, diffract, or reflect a blackbody source obliquely off of a dielectric or metal. Such an effect is observed in the case of sunlight reflecting off the surface of a body of water or the glass of an automobile.

In terms of the linear orthogonal polarization states, and assuming propagation in the z-direction and detection at the origin, a plane wave can be expressed as,

$$E_x(t) = \int_{-\infty}^{\infty} \mathcal{E}_x(\omega, t)\mathrm{e}^{\mathrm{i}\phi_x(\omega, t)}\mathrm{e}^{\mathrm{i}k(\omega)z - \mathrm{i}\omega t}d\omega, \tag{2.12}$$

$$E_y(t) = \int_{-\infty}^{\infty} \mathcal{E}_y(\omega, t)\mathrm{e}^{\mathrm{i}\phi_y(\omega, t)}\mathrm{e}^{\mathrm{i}k(\omega)z - \mathrm{i}\omega t}d\omega. \tag{2.13}$$

Here, the amplitudes of the x- and y-components of the radiation electric field spectral density as a function of frequency and time are $\mathcal{E}_x(\omega, t)$ and $\mathcal{E}_y(\omega, t)$. The phases are in general random and represented by $\phi_x(\omega, t)$ and $\phi_y(\omega, t)$.

Using these expressions, it is possible to represent unpolarized light. $E_x(t)$ and $E_y(t)$ are not necessarily coherent, and the degree to which partial polarization occurs is a function of the nature of the coherence of these functions.

Returning briefly to the above discussion on modification of the polarization state of light, for natural light, in addition to polarization-preserving polarization transformations, polarization and depolarization can take place by altering the state of *coherence* of the radiation. For example, one can polarize a signal by increasing the coherence (e.g., by using a polarizing analyzer to separate fields by polarization state) or depolarize a signal by decreasing its coherence (e.g., through the use of a diffuser to randomize the field alignment).

2.6 Electromagnetic Coherence

For the general case of partially-coherent light the coherence can be characterized as a function of spatial and temporal scale. Light from random emitters has a finite frequency range and phase fluctuations that cause the coherence to become "washed out" over distance and time. However, events along the path of the light can alter the correlations present in the light. For example, scattering from a spatially-coherent structure can increase the level of correlation while passage through time-varying media (e.g., turbulence) can lead to a decrease. For a finite bandwidth $\Delta\nu$, we would expect that for times smaller than $1/\Delta\nu$, the waves superpose in a way that essentially behaves like a single-frequency wave. The coherence time is then defined as

$$\Delta\tau_c \approx \frac{1}{\Delta\nu}. \tag{2.14}$$

There is a corresponding coherence length that is related to the coherence time by the speed of light, c,

$$\Delta\ell_c \approx c\Delta\tau_c. \tag{2.15}$$

The coherence of natural light can be quantified by the cross-correlation of two quantities, taken at zero time delay. We define two functions, $\xi(t)$ and $\psi(t)$ and define their *correlation* over a time interval T,

$$\langle \xi(t)^* \psi(t + \tau) \rangle_T \equiv \int_T \xi(t)^* \psi(t + \tau) dt. \tag{2.16}$$

For polarization, we are particularly interested in correlating functions with $\tau = 0$. If $\xi(t)$ and $\psi(t)$ are uncorrelated, that is, their temporal fluctuations are statistically independent of one another, then on average over a sufficiently long period of time, the number of positive contributions to the integral (i.e., when both functions fluctuate together and the product is positive) will cancel the number of negative contributions (i.e., when the functions fluctuate in opposite directions, and the product is negative). The result is that this integral on average tends to zero for uncorrelated functions. If on the other hand, the functions are correlated, they will fluctuate together more than in opposite directions, and the integral will be positive. If the two functions are anticorrelated, they will fluctuate in opposite directions more than in the same direction, and the integral will be negative. Polarization is a manifestation of the correlations between vector components of the electric field.

2.7 Polarization as a Statistical Quantity

The Jones vector formalism is excellent for handling monochromatic light, i.e., in settings where the magnitudes of each orthogonal polarization and the relative phase between them fully characterizes the polarization state. As a reminder, monochromatic light is always 100% polarized because electric fields add coherently for this case. However, we are also now familiar with the fact that there exists unpolarized and partially polarized light. So, how does this happen?

Real sources emit light for finite amounts of time and thus have a finite spectral bandwidth. In addition, the emission phase is largely random for many astronomical radiation sources. These factors lead to the production of *partially-coherent* light. This is a very general case, and physical processes involving anisotropy can conspire to generate some coherence (even in the case of blackbody radiation—the most incoherent light-producing physical process that we know of). In this case, it does not make sense to discuss the electric field value as a function of time, because of the random aspect of its phase. Instead, it is of value to consider time averages (denoted by angled brackets below). Generally, we can consider the matrix of correlations of the transverse components of the radiation field between any two points, $\mathbf{r}_1$ and $\mathbf{r}_2$ over some time delay τ. This is called the *electrical mutual coherence matrix*,

$$\Gamma(\mathbf{r}_1, \mathbf{r}_2, \tau) = \begin{pmatrix} \langle E_x^*(\mathbf{r}_1, t) E_x(\mathbf{r}_2, t + \tau) \rangle & \left\langle E_y^*(\mathbf{r}_1, t) E_x(\mathbf{r}_2, t + \tau) \right\rangle \\ \left\langle E_x^*(\mathbf{r}_1, t) E_y(\mathbf{r}_2, t + \tau) \right\rangle & \left\langle E_y^*(\mathbf{r}_1, t) E_y(\mathbf{r}_2, t + \tau) \right\rangle \end{pmatrix}. \tag{2.17}$$

Here the (*) denotes complex conjugation.

Another important quantity related to the mutual coherence matrix is the *cross-spectral density matrix*,

$$\begin{pmatrix} \left\langle \tilde{E}_x^*(\mathbf{r}_1, \omega)\tilde{E}_x(\mathbf{r}_2, \omega)\right\rangle & \left\langle \tilde{E}_y^*(\mathbf{r}_1, \omega)\tilde{E}_x(\mathbf{r}_2, \omega)\right\rangle \\ \left\langle \tilde{E}_x^*(\mathbf{r}_1, \omega)\tilde{E}_y(\mathbf{r}_2, \omega)\right\rangle & \left\langle \tilde{E}_y^*(\mathbf{r}_1, \omega)\tilde{E}_y(\mathbf{r}_2, \omega)\right\rangle \end{pmatrix}. \tag{2.18}$$

Here, the brackets correspond to the ensemble averages rather than time average. Note that when $\mathbf{r}_1 = \mathbf{r}_2$, the diagonal elements are the power spectral densities of the two orthogonal polarizations.

> The cross-spectral density is the Fourier transform of the electric mutual coherence matrix. This is a vector extension to the *Wiener–Khinchin Theorem* that is revisited and derived in the context of signal processing theory for scalar quantities in Chapter 6.

For polarization analysis, we are interested in the specific correlations at a given point in space and moment in time. Setting $\mathbf{r}_1 = \mathbf{r}_2$ gives a specific "slice" of the mutual coherence matrix (or cross-spectral density) that completely describes the polarization,

$$\begin{pmatrix} \langle E_x^2\rangle & \left\langle E_x E_y^*\right\rangle \\ \left\langle E_x^* E_y\right\rangle & \left\langle E_y^2\right\rangle \end{pmatrix}. \tag{2.19}$$

From the Wiener–Khinchin theorem, we can choose to consider the problem from the frequency domain, in which the angle brackets represent an ensemble average, and we have integrated over the bandpass of the instrument. Motivated by this perspective, the matrix is often called the *density matrix*. When considered in the time domain, the angle brackets are the time average over the integration time of the detector. In the time domain, this matrix is called the *coherency* matrix. In the limit these signals can be treated as ergodic, the two pictures are equivalent. In the nomenclature of Wolf (2007) this is designated as the *polarization matrix*.

Jones matrices can be utilized in a partially-coherent system. However, instead of using them to operate on Jones vectors that characterize the exact phase relationship between the two components of the electric field, the polarization matrix, which is statistical in nature, can be constructed from Jones vectors, which contain complex field amplitudes for two orthogonal (and transverse) components. The polarization matrix can be composed of the *outer product* of a Jones vector with itself, followed by a time average,

$$\mathbf{D} \equiv \langle |E\rangle\langle E|\rangle = \left\langle \begin{pmatrix} E_x \\ E_y \end{pmatrix} \begin{pmatrix} E_x^* & E_y^* \end{pmatrix} \right\rangle = \begin{pmatrix} \langle E_x^2\rangle & \left\langle E_x E_y^*\right\rangle \\ \left\langle E_y E_x^*\right\rangle & \left\langle E_y^2\right\rangle \end{pmatrix}.$$

In the case of the polarization matrix, the correlations are considered between orthogonal polarizations at a single point in space. The more general correlation matrix also considers spatial correlations in the radiation field. The transformation of the polarization matrix follows from the transformation of the Jones vectors,

$$\mathbf{D}' = \mathbf{J}\mathbf{D}\mathbf{J}^{\dagger} = \langle \mathbf{J}|E\rangle\langle E|\mathbf{J}^{\dagger}\rangle.$$

Here, (†) is the Hermitian conjugate.

2.8 Stokes' Parameters

The polarization matrix completely describes the polarization state of partially-coherent light, but is not particularly physically illuminating. The diagonal terms are just the amplitudes for the two polarizations; the off-diagonal elements encode the correlations between the x- and y- components. More insight can be gained by writing this matrix as a linear combination of the identity matrix and the Pauli spin matrices,

$$I\begin{pmatrix}1 & 0\\ 0 & 1\end{pmatrix} + Q\begin{pmatrix}1 & 0\\ 0 & -1\end{pmatrix} + U\begin{pmatrix}0 & 1\\ 1 & 0\end{pmatrix} + V\begin{pmatrix}0 & -i\\ i & 0\end{pmatrix}. \tag{2.20}$$

This leads to the following definitions for the coefficients above:

$$I = \langle E_x^2\rangle + \langle E_y^2\rangle \tag{2.21}$$

$$Q = \langle E_x^2\rangle - \langle E_y^2\rangle \tag{2.22}$$

$$U = \langle E_x^* E_y\rangle + \langle E_x E_y^*\rangle = 2\Re\langle E_x^* E_y\rangle \tag{2.23}$$

$$V = \mathrm{i}\left(\langle E_x^* E_y\rangle - \langle E_x E_y^*\rangle\right) = 2\Im\langle E_x^* E_y\rangle. \tag{2.24}$$

These are the *Stokes Parameters*. Stokes I is simply (proportional to) the total power. Stokes Q is the difference in power between the horizontal and vertical polarization states. Stokes U is the same quantity in a basis that is oriented at an angle of 45° relative to both axes of the original basis. This can be understood as the "in-phase" coherence between the horizontal and vertical polarizations. In other words, if two orthogonal vectors are added, they produce a vector that is rotated by 45°. Stokes V is the degree of circular polarization. In analogy to U, this is the "quadrature" coherence between the x and y components, and thus will be maximal when the phase between the two is 90°.

In terms of the Jones vector of a system, the Stokes parameters can be expressed using the Pauli matrices,

$$I = \langle E|\sigma_0|E\rangle, \tag{2.25}$$

$$Q = \langle E|\sigma_1|E\rangle, \tag{2.26}$$

$$U = \langle E|\sigma_2|E\rangle, \tag{2.27}$$

$$V = \langle E|\sigma_3|E\rangle. \tag{2.28}$$

It is straightforward to calculate the polarization matrix in terms of the Stokes parameters,

$$\mathbf{D} = \frac{1}{2}\begin{pmatrix} I + Q & U - \mathrm{i}V \\ U - \mathrm{i}V & I - Q \end{pmatrix}. \tag{2.29}$$

This expression is useful if it is desirable to convert measured Stokes parameters to the elements of the polarization matrix. The reverse operation, that is obtaining the Stokes parameters from the polarization matrix, can be obtained by

$$I = \mathrm{tr}(\mathbf{D}\sigma_0), \tag{2.30}$$

$$Q = \mathrm{tr}(\mathbf{D}\sigma_1), \tag{2.31}$$

$$U = \mathrm{tr}(\mathbf{D}\sigma_2), \tag{2.32}$$

$$V = \mathrm{tr}(\mathbf{D}\sigma_3). \tag{2.33}$$

It is also useful to find the determinant and trace of the polarization matrix.

$$\det(\mathbf{D}) = \frac{1}{4}(I^2 - Q^2 - U^2 - V^2) \equiv \frac{1}{4}(I^2 - P^2), \tag{2.34}$$

$$\mathrm{tr}(\mathbf{D}) = I. \tag{2.35}$$

Here, we have defined the total polarization P, which obeys the inequality

$$I^2 \geqslant P^2 \equiv Q^2 + U^2 + V^2. \tag{2.36}$$

The equality holds for 100% polarized light. The four Stokes parameters completely describe the state of coherence of general partially-coherent light. Note that only three parameters are required to describe the state of monochromatic light, whereas the four Stokes parameters are required to describe partially-polarized light.

2.9 Normalized Stokes Parameters and Fractional Polarization

Often in astronomical polarization, the fractional amount of light that is polarized is the parameter of interest in the scientific interpretations of the data. In this case it is useful to define the normalized Stokes parameters,

$$q \equiv \frac{Q}{I}, \quad u \equiv \frac{U}{I}, \quad v \equiv \frac{V}{I}. \tag{2.37}$$

Given these quantities, the *fractional polarization* (sometimes referred to as *degree of polarization*), p, and the *polarization angle* (also known as the *position angle*), ϕ, as

$$p = \sqrt{q^2 + u^2 + v^2} = \sqrt{\frac{Q^2 + U^2 + V^2}{I^2}}, \tag{2.38}$$

$$\phi = \frac{1}{2}\arctan\frac{u}{q}. \tag{2.39}$$

The fractional polarization can be derived directly from the polarization matrix (Brosseau 1998),

$$p^2 = 1 - 4\frac{\det(\mathbf{D})}{[\mathrm{tr}(\mathbf{D})]^2}. \tag{2.40}$$

Because the trace and determinant are rotational invariants, this expression displays the property that the polarization is also a rotational invariant. That is, it does not depend on the basis chosen to express the correlations relevant to polarization. Note that the trace of the polarization matrix is simply Stokes I.

2.10 Müller Matrix Techniques

The Stokes parameters represent real quantities that have the units of power (or intensity or flux depending on the measurement). These quantities, unlike the elements of the polarization matrix, are directly measurable. Thus, for measurement purposes, it is often useful to recast the system response in terms of its effect on the Stokes parameters directly. In this case, one defines a *Stokes vector* as a real 1×4 column vector,

$$\mathbf{s} = \begin{pmatrix} I \\ Q \\ U \\ V \end{pmatrix}. \tag{2.41}$$

Transformations in the Stokes vector space can be represented by a group of 4×4 matrices that act on a Stokes vector. These are called Müller matrices. The transformation of an optical system can be written in terms of its effects on the Stokes parameters,

$$\begin{pmatrix} I' \\ Q' \\ U' \\ V' \end{pmatrix} = \begin{pmatrix} M_{II} & M_{IQ} & M_{IU} & M_{IV} \\ M_{QI} & M_{QQ} & M_{QU} & M_{QV} \\ M_{UI} & M_{UQ} & M_{UU} & M_{UV} \\ M_{VI} & M_{VQ} & M_{VU} & M_{VV} \end{pmatrix} \begin{pmatrix} I \\ Q \\ U \\ V \end{pmatrix}. \tag{2.42}$$

This linear transformations is commonly written as $\mathbf{s}' = \mathbf{M}\mathbf{s}$.

For modeling a system with known components, knowledge of the propagation phase is usually important. For systems for which the length scales are of the same order as the coherence length or smaller, boundary conditions can couple the effects of multiple elements by reprocessing the light. In these cases, the phase of each mode must be tracked and response of elements cannot be treated incoherently. On the other hand, in systems where the distances are much larger than the coherence length, each element may be treated independently, and the incoherent approach of the Müller matrices is sufficient.

2.11 The Poincaré Sphere

The polarization state can be visualized by mapping the q, u, and v values to the three-dimensional Cartesian axes, as shown in Figure 2.3. Typically, the q and u axes are mapped to the horizontal plane. Points in this plane represent linear polarized states. Spheres that are concentric with the origin (each is known as a *Poincaré* Sphere) represent the set of states with constant total polarization $p = \sqrt{q^2 + u^2 + v^2}$. The upper pole of the Poincaré sphere represents right-handed polarization, and the bottom pole of this space represents left-handed polarization States between the plane and the poles are generalized elliptical polarization states. The space is bounded by the $p = 1$ surface, since $p \leqslant 1$.

The azimuthal angle on the Poincaré sphere (2ϕ) is twice the physical position angle. This gives the direction of the "vector" for linear polarization; for elliptical polarization, this is the orientation of the semimajor axis of the ellipse traced out by the polarization. The angle δ_0, the complement of the polar angle, represents the phase delay between two orthogonal linear polarizations in the linear basis where the electric field has equal magnitude in each direction.

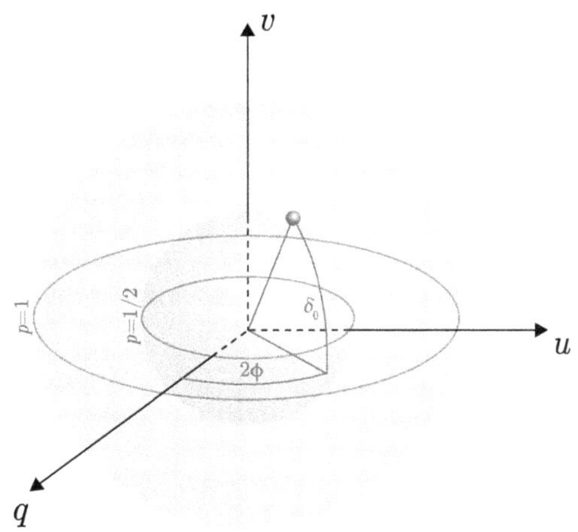

Figure 2.3. In a space where the normalized Stokes parameters are identified with the Cartesian axes, spheres centered at the origin represented surfaces for which the polarization fraction $p = \sqrt{q^2 + u^2 + v^2}$ = constant is the radius. Surfaces containing the states $p = 1/2$ and $p = 1$ are explicitly shown. The polarization fraction, p, is bounded between the values of 0 for unpolarized light and 1 for completely polarized light. The angle ϕ represents the angle of the linear polarization, for states in the q–u plane and the angle of the semimajor axis for general elliptical states. The angle δ_0 represents the degree of circularity of the polarization.

In this Stokes vector space, physical processes that change the state of polarization while keeping the total polarization transform the state to other states that lie on the Poincaré sphere, a surface for which p=constant. This is the case for processes that do not affect the amplitudes of the polarization, but introduce a coherent phase shift between orthogonal polarization states. One can utilize either the Stokes parameters or the normalized Stokes parameters. Here we choose the latter, so that the radius of the spherical surfaces represent the set of states having a constant p. For example, $p = 1$ represents the spherical surface containing all of the states that are 100% polarized.

The Poincaré sphere provides a geometric visualization of transformations represented by Müller matrices. On the Poincaré sphere, orthogonal polarization states lie at opposite ends of any line segment through the center of the sphere. Phase delays between any two orthogonal states cause a polarization transformation, represented graphically as a rotation about the axis connecting the two orthogonal states that has the same angular value as the phase delay itself (Tinbergen 1996). For example, we may be interested specifically in a polarization transformation resulting from a phase delay introduced between orthogonal circular polarization states. This results in a rotation about the v-axis of the Poincaré sphere (see Figure 2.3). Phase delays between two orthogonal polarizations uniquely describe operations that preserve the overall state of polarization, that is p (or P). These are described by great circle paths on the surface of a given sphere in this Poincaré space.

The corresponding Müller matrices to these transformations are orthogonal matrices in the three polarized dimensions (Q, U, V). All such matrices constitute a representation of the group $O(3)$. There is a homomorphism between $O(3)$ and the group $SU(2)$, the special unitary group of order 2. The Jones matrices for the same set of physical transformations can be used as a representation of $SU(2)$ (Sternberg 1994). The upshot here is that the polarization-preserving operations on a beam in an optical system, which are uniquely realized by the introduction of a phase delay between orthogonal polarizations, form a group, and thus we can infer some properties about polarization transformations based on the general properties of a group.

- The identity operation corresponds to the addition of a phase delay of zero.
- Each operation has an inverse that corresponds to the introduction of a phase of the opposite sign.
- Two successive phase delays can be mapped to an equivalent single phase delay (between appropriate orthogonal polarizations).

In practice, for beams of finite bandwidth, depolarization occurs in polarization transformation, because for many of the techniques used to introduce the phase, the magnitude of the delay is wavelength-dependent. Considering polarimetric transformations in terms of group theory is useful in the development of broadband polarization transformers, as we will see later in this book. In the next section, we will show explicitly the relationship between a phase delay and polarization transformation, along with a demonstration of converting between the Jones and Müller formalisms.

2.12 Example: Introduction of a Phase Delay Between Linear Polarizations

Let's consider the transformation of a set of Stokes parameters by the introduction of a phase delay, δ, between orthogonal linear polarizations. We can represent the phase delay by a Jones matrix. For symmetry, we'll advance the x-component by half the phase and delay the y-component by half the phase,

$$\begin{aligned}\mathbf{J} &= \begin{pmatrix} e^{i\delta/2} & 0 \\ 0 & e^{-i\delta/2} \end{pmatrix} \\ \mathbf{J}^\dagger &= \begin{pmatrix} e^{-i\delta/2} & 0 \\ 0 & e^{i\delta/2} \end{pmatrix}.\end{aligned}$$

We can then transform the general density matrix,

$$\mathbf{D}' = \mathbf{JDJ}^\dagger = \mathbf{J}(I\boldsymbol{\sigma}_0 + Q\boldsymbol{\sigma}_1 + U\boldsymbol{\sigma}_2 + V\boldsymbol{\sigma}_3)\mathbf{J}^\dagger.$$

We can certainly perform the matrix multiplications longhand, but we will instead utilize the Pauli algebra. We know that the following matrix multiplications are true, and these can be used to organize our calculations,

$$\begin{aligned}\boldsymbol{\sigma}_i\boldsymbol{\sigma}_j &= i\varepsilon_{ijk}\boldsymbol{\sigma}_k (i \neq j) \\ \boldsymbol{\sigma}_0\boldsymbol{\sigma}_i &= \boldsymbol{\sigma}_i\boldsymbol{\sigma}_0 = \boldsymbol{\sigma}_i \\ \boldsymbol{\sigma}_i\boldsymbol{\sigma}_i &= \boldsymbol{\sigma}_0.\end{aligned}$$

We next express the Jones matrix as a linear combination of Pauli matrices,

$$\begin{aligned}\begin{pmatrix} e^{i\delta/2} & 0 \\ 0 & e^{-i\delta/2} \end{pmatrix} &= A\begin{pmatrix} 1 & 0 \\ 0 & 1 \end{pmatrix} + B\begin{pmatrix} 1 & 0 \\ 0 & -1 \end{pmatrix} \\ A + B &= e^{i\delta/2}, \quad A - B = e^{-i\delta/2} \\ 2A &= e^{i\delta/2} + e^{-i\delta/2} \rightarrow A = \cos\delta/2 \\ 2B &= e^{i\delta/2} - e^{-i\delta/2} \rightarrow B = i\sin\delta/2.\end{aligned}$$

Thus, we have

$$\begin{aligned}\mathbf{J} &= \boldsymbol{\sigma}_0\cos\delta/2 + i\boldsymbol{\sigma}_1\sin\delta/2 \\ \mathbf{J}^\dagger &= \boldsymbol{\sigma}_0\cos\delta/2 - i\boldsymbol{\sigma}_1\sin\delta/2.\end{aligned}$$

The general transformation can be expressed,

$$\mathbf{D}' = (\boldsymbol{\sigma}_0\cos\delta/2 + i\boldsymbol{\sigma}_1\sin\delta/2)(I\boldsymbol{\sigma}_0 + Q\boldsymbol{\sigma}_1 + U\boldsymbol{\sigma}_2 + V\boldsymbol{\sigma}_3)(\boldsymbol{\sigma}_0\cos\delta/2 - i\boldsymbol{\sigma}_1\sin\delta/2).$$

This is just an algebraic expression. Once simplified, we will look for the coefficients, which are the new values of the Stokes parameters relative to the initial values (I, Q, U, V). To organize the calculation, we can examine how each of these terms transform:

The transformation of I:

$$
\begin{aligned}
&(\boldsymbol{\sigma}_0 \cos\delta/2 + \mathrm{i}\boldsymbol{\sigma}_1 \sin\delta/2) I \boldsymbol{\sigma}_0 (\boldsymbol{\sigma}_0 \cos\delta/2 - \mathrm{i}\boldsymbol{\sigma}_1 \sin\delta/2)\\
&= \boldsymbol{\sigma}_0\boldsymbol{\sigma}_0\boldsymbol{\sigma}_0 I \cos^2\delta/2 - \boldsymbol{\sigma}_0\boldsymbol{\sigma}_0\boldsymbol{\sigma}_1 \mathrm{i} I \cos\delta/2 \sin\delta/2 + \boldsymbol{\sigma}_1\boldsymbol{\sigma}_0\boldsymbol{\sigma}_0 \mathrm{i} I \sin\delta/2 \cos\delta/2\\
&+ \boldsymbol{\sigma}_1\boldsymbol{\sigma}_0\boldsymbol{\sigma}_1 I \sin^2\delta/2\\
&= \boldsymbol{\sigma}_0 I(\cos^2\delta/2 + \sin^2\delta/2) + \boldsymbol{\sigma}_1 \mathrm{i} I \cos\delta/2 \sin\delta/2(1 - 1)\\
&= I\boldsymbol{\sigma}_0.
\end{aligned}
$$

Thus, there is no change in the total intensity.

The transformation of Q:

$$
\begin{aligned}
&(\boldsymbol{\sigma}_0 \cos\delta/2 + \mathrm{i}\boldsymbol{\sigma}_1 \sin\delta/2) Q \boldsymbol{\sigma}_1 (\boldsymbol{\sigma}_0 \cos\delta/2 - \mathrm{i}\boldsymbol{\sigma}_1 \sin\delta/2)\\
&= \boldsymbol{\sigma}_0\boldsymbol{\sigma}_1\boldsymbol{\sigma}_0 Q \cos^2\delta/2 - \boldsymbol{\sigma}_0\boldsymbol{\sigma}_1\boldsymbol{\sigma}_1 \mathrm{i} Q \cos\delta/2 \sin\delta/2 + \boldsymbol{\sigma}_1\boldsymbol{\sigma}_1\boldsymbol{\sigma}_0 \mathrm{i} Q \sin\delta/2 \cos\delta/2\\
&+ \boldsymbol{\sigma}_1\boldsymbol{\sigma}_1\boldsymbol{\sigma}_1 Q \sin^2\delta/2\\
&= \boldsymbol{\sigma}_0 Q(\cos^2\delta/2 + \sin^2\delta/2) + \boldsymbol{\sigma}_1 \mathrm{i} Q \cos\delta/2 \sin\delta/2(1 - 1)\\
&= Q\boldsymbol{\sigma}_0.
\end{aligned}
$$

There is no change in Q either.

The transformation of U:

$$
\begin{aligned}
&(\boldsymbol{\sigma}_0 \cos\delta/2 + \mathrm{i}\boldsymbol{\sigma}_1 \sin\delta/2) U \boldsymbol{\sigma}_2 (\boldsymbol{\sigma}_0 \cos\delta/2 - \mathrm{i}\boldsymbol{\sigma}_1 \sin\delta/2)\\
&= \boldsymbol{\sigma}_0\boldsymbol{\sigma}_2\boldsymbol{\sigma}_0 U \cos^2\delta/2 - \boldsymbol{\sigma}_0\boldsymbol{\sigma}_2\boldsymbol{\sigma}_1 \mathrm{i} U \cos\delta/2 \sin\delta/2 + \boldsymbol{\sigma}_1\boldsymbol{\sigma}_2\boldsymbol{\sigma}_0 \mathrm{i} U \sin\delta/2 \cos\delta/2\\
&+ \boldsymbol{\sigma}_1\boldsymbol{\sigma}_2\boldsymbol{\sigma}_1 U \sin^2\delta/2\\
&= \boldsymbol{\sigma}_2 U \cos^2\delta/2 - \boldsymbol{\sigma}_3 U \cos\delta/2 \sin\delta/2 - \boldsymbol{\sigma}_3 U \cos\delta/2 \sin\delta/2 - \boldsymbol{\sigma}_2 U \sin^2\delta/2\\
&= U(\boldsymbol{\sigma}_2 \cos\delta - \boldsymbol{\sigma}_3 \sin\delta).
\end{aligned}
$$

The transformation of V:

$$
\begin{aligned}
&(\boldsymbol{\sigma}_0 \cos\delta/2 + \mathrm{i}\boldsymbol{\sigma}_1 \sin\delta/2) V \boldsymbol{\sigma}_3 (\boldsymbol{\sigma}_0 \cos\delta/2 - \mathrm{i}\boldsymbol{\sigma}_1 \sin\delta/2)\\
&= \boldsymbol{\sigma}_0\boldsymbol{\sigma}_3\boldsymbol{\sigma}_0 V \cos^2\delta/2 - \boldsymbol{\sigma}_0\boldsymbol{\sigma}_3\boldsymbol{\sigma}_1 \mathrm{i} V \cos\delta/2 \sin\delta/2 + \boldsymbol{\sigma}_1\boldsymbol{\sigma}_3\boldsymbol{\sigma}_0 \mathrm{i} V \sin\delta/2 \cos\delta/2\\
&+ \boldsymbol{\sigma}_1\boldsymbol{\sigma}_3\boldsymbol{\sigma}_1 V \sin^2\delta/2\\
&= \boldsymbol{\sigma}_3 V \cos^2\delta/2 + \boldsymbol{\sigma}_2 V \cos\delta/2 \sin\delta/2 + \boldsymbol{\sigma}_2 V \cos\delta/2 \sin\delta/2 - \boldsymbol{\sigma}_2 V \sin^2\delta/2\\
&= V(\boldsymbol{\sigma}_3 \cos\delta + \boldsymbol{\sigma}_2 \sin\delta).
\end{aligned}
$$

We now put these together into the transformation,

$$
\mathbf{D}' = I\boldsymbol{\sigma}_0 + Q\boldsymbol{\sigma}_1 + (U\cos\phi + V\sin\delta)\boldsymbol{\sigma}_2 + (V\cos\delta - U\sin\delta)\boldsymbol{\sigma}_3. \tag{2.43}
$$

The contribution to the transformed polarization matrix from each Stokes parameter is the coefficient of the respective matrix. We can solve for the new Stokes parameters in terms of the old ones via proper identification of the coefficients in Equation (2.43),

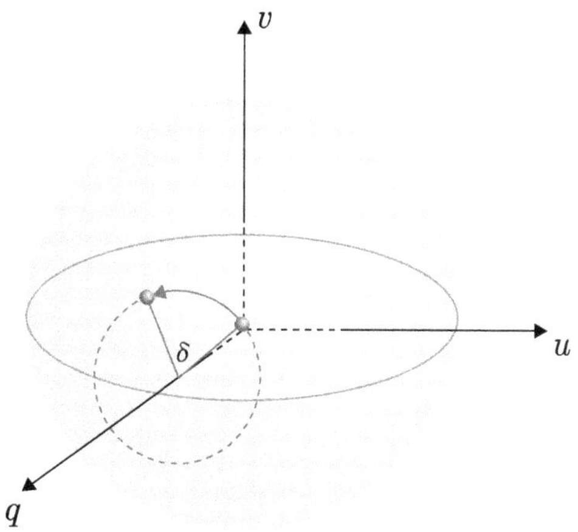

Figure 2.4. The transformation of a phase delay maps onto a sphere as a rotation around the line between the two orthogonal polarizations between which the phase delay δ is introduced. In this case, the phase delay is introduced between horizontal (q) and vertical ($-q$) polarization states.

$$
\begin{aligned}
I' &= I \\
Q' &= Q \\
U' &= U\cos\delta + V\sin\delta \\
V' &= V\cos\delta - U\sin\delta.
\end{aligned}
$$

These expressions directly enable the writing of a Müller matrix that represents this phase delay,

$$
\begin{aligned}
\begin{pmatrix} I' \\ Q' \\ U' \\ V' \end{pmatrix} &= \begin{pmatrix} 1 & 0 & 0 & 0 \\ 0 & 1 & 0 & 0 \\ 0 & 0 & \cos\delta & \sin\delta \\ 0 & 0 & -\sin\delta & \cos\delta \end{pmatrix} \begin{pmatrix} I \\ Q \\ U \\ V \end{pmatrix} \\
\mathbf{s}' &= \mathbf{Ms}.
\end{aligned}
$$

The transformation is shown on the Poincaré sphere in Figure 2.4. The phase delay is manifest as a rotation about the line connecting two orthogonal polarization states between which the phase is introduced. The rotation angle is equal to the phase delay. In this case, the polarization states between which the phase is introduced between the horizontal (represented by the state on the q axis) and vertical (represented by the state on the $-q$ axis.) The total polarization does not change; the initial and final states lie on the same Poincaré sphere.

2.13 Comparing Jones and Müller Formalisms

The Müller matrix formalism encodes the transformation of the coherence state of radiation by a system. The effect of a Müller matrix on a state vector (which consists of the four Stokes parameters) tells us very little about the nature of the system intrinsically, since phase information is only implicitly included via the transformations of the correlations among and between fields.

A Comment on Matrix Formalisms

There is a nice parallel between the relationship between Jones and Müller matrices in polarization analysis and the use of matrices in microwave engineering (Pozar 2012).

In transmission line theory, detailed circuit analysis and design is commonly performed using so-called "ABCD" matrices. These matrices take into account the boundary conditions and phases of the complex voltages and currents encountered in computing the response of a cascaded circuit network. However, in validating the circuit response the complex *scattering parameter* amplitudes are characterized (i.e., with vector network analysis techniques). To directly compare and contrast theory to observation a transformation between ABCD and scattering matrix parameterizations is necessary. In going beyond the scale of the field coherence length, the power reflectance, transmittance, or scattering cross section are typically characterized (i.e., with scalar measurement techniques). These basic frameworks for coherent ("vector" or complex amplitudes) and incoherent ("scalar" or power) network analysis and measurement can be generalized to handle multiple ports and/or modes.

Analogous considerations arise in polarization-sensitive systems, which represent an example of a dual-mode network. Jones matrices are utilized in parts of the optical system that are sensitive to coherence and polarization coupling or recirculated light between elements if present. However, in going beyond the coherence length scale, individual fields are not readily observable. Rather, the Stokes parameters represent integrated quantities in power units that can be readily measured and used to characterize the system response. In this limit, Müller matrices can be used to represent transformations that capture the statistical relationships of the phases between orthogonal polarizations.

References

Balanis, C. A. 2024, Balanis' Advanced Engineering Electromagnetics (New York: Wiley)

Brosseau, C. 1998, Fundamentals of Polarized (1st ed.; New York: Wiley)

Goldstein, D. 2003, Polarized Light (New York: Dekker)

Górski, K. M., Hivon, E., Banday, A. J., et al. 2005, ApJ, 622, 759

Harrington, R. F. 1987, JEWA, 1, 181

Harrington, R. F. 1993, Field Computation by Moment Methods (Piscataway, NJ: Wiley-IEEE Press)

Hecht, E. 2017, Optics (5th ed.; New York: Pearson)

Jin, J.-M. 2015, The finite element method in electromagnetics (New York: Wiley)

Jones, R. C. 1941, JOSA, 31, 488

Pozar, D. M. 2012, Microwave Engineering (4th ed.; Hoboken, NJ: Wiley)

Sternberg, S. 1994, Group Theory and Physics (1st ed.; Cambridge: Cambridge Univ. Press)

Tinbergen, J. 1996, Astronomical Polarimetry (1st ed.; Cambridge: Cambridge Univ. Press)

Wiener, O. 1890, AnP, 276, 203

Wolf, E. 2007, Introduction to the Theory of Coherence and Polarization of Light (1st ed.; Cambridge: Cambridge Univ. Press)

Yee, K. 1966, ITAP, 14, 302

Chapter 3

Polarizing Processes

As mentioned in the introduction, polarization results from an asymmetry (or anisotropy) in processes affecting the propagation of electromagnetic radiation. These asymmetries lead to correlations between field components that are manifest in the observed Stokes parameters. In this chapter, we review select physical processes that are particularly relevant to astrophysical polarization as well as instrument design considerations. To motivate how polarization is produced we start with a general review of how light interacts with insulating and conductive media. Examples that are potentially useful when considering the design of instrumentation for polarization measurement are also touched upon.

3.1 Maxwell's Equations in General Media

To consider the use of materials in polarization measurement, we begin with the general form of Maxwell's equations in a material,

$$\begin{aligned} \nabla \cdot \mathbf{D} &= \rho \qquad \nabla \times \mathbf{E} = -\frac{\partial \mathbf{B}}{\partial t} \\ \nabla \cdot \mathbf{B} &= 0 \qquad \nabla \times \mathbf{H} = \mathbf{J} + \frac{\partial \mathbf{D}}{\partial t}. \end{aligned} \tag{3.1}$$

Here, we keep in mind the constitutive relations,

$$\mathbf{B} = \mu \mathbf{H} \tag{3.2}$$

$$\mathbf{D} = \varepsilon \mathbf{E}, \tag{3.3}$$

where ε and μ are the permittivity and permeability that describe the electromagnetic energy stored in the material. For conductive materials, the current density is given by $\mathbf{J} = \sigma \mathbf{E}$, where σ is the bulk conductivity of the medium. In this book, we will focus on non-magnetic materials for which $\mu = \mu_{\rm o}$.

doi:10.1088/2514-3433/ae1022ch3

We consider the fields in a highly conducting medium according to Maxwell's equations, for which we assume the material has no free charges ($\rho = 0$) but can have a non-zero current density $\mathbf{J}$. We start with the laws of Faraday and Ampere,

$$\nabla \times \mathbf{E} = \mu \frac{\partial \mathbf{H}}{\partial t}, \tag{3.4}$$

$$\nabla \times \mathbf{H} = \mathbf{J} + \varepsilon \frac{\partial \mathbf{E}}{\partial t}. \tag{3.5}$$

Taking the curl of the first equation (Faraday's law) yields

$$\nabla \times \nabla \times \mathbf{E} = \mu \frac{\partial}{\partial t}(\nabla \times \mathbf{H}), \tag{3.6}$$

$$-\nabla^2 \mathbf{E} = \mu \frac{\partial \mathbf{J}}{\partial t} + \mu\varepsilon \frac{\partial^2 \mathbf{E}}{\partial t^2}. \tag{3.7}$$

For simplicity we assume that the material is Ohmic and the current density is related to the electric field using a frequency-dependent conductivity $\sigma(\omega)$, $\mathbf{J} = \sigma(\omega)\mathbf{E}$,

$$-\nabla^2 \mathbf{E} = \mu\sigma \frac{\partial \mathbf{E}}{\partial t} + \mu\varepsilon \frac{\partial^2 \mathbf{E}}{\partial t^2}. \tag{3.8}$$

We assume a plane wave solution for the electric field vector that can be written in terms of its harmonic components, $\mathbf{E} = \mathbf{E}_0 e^{i(\mathbf{k}\cdot\mathbf{r} - \omega t)}$, where $k = \tilde{n}k_o$. $\tilde{n}$ is the index of refraction in the medium and $k_o = 2\pi/\lambda_o$ is the magnitude of the wavevector in free space,

$$k^2\mathbf{E} = \mu(i\sigma\omega + \varepsilon\omega^2)\mathbf{E} = \mu(i\sigma/\omega + \varepsilon)\omega^2\mathbf{E} = \mu\hat{\varepsilon}\omega^2\mathbf{E}. \tag{3.9}$$

We can define an effective permittivity that has contributions from stored energy and conduction through the medium,

$$\hat{\varepsilon} = \varepsilon + i\frac{\sigma}{\omega}. \tag{3.10}$$

The first term arises from the energy stored in the displacement field; the second comes from resistive loss in the material. We can also write a dispersion relation based on Equation (3.9),

$$\frac{\omega^2}{k^2} = 1/\mu\hat{\varepsilon} = c^2/\tilde{n}^2. \tag{3.11}$$

Here, c is the speed of light in a vacuum, and $\tilde{n}$ is the complex index of refraction.

3.2 Drude Dielectric Model

When thinking about a particular medium, be it a metal or the ISM, we need a model for the conductivity. The Drude model for metals, which assumes that the conduction electrons are free and interact with atoms (Ashcroft & Mermin 1976) yields a classical optical conductivity of

$$\sigma(\omega) = \frac{\sigma_o}{1 - i\omega\tau}, \tag{3.12}$$

where

$$\sigma_o = \frac{n_e e^2 \tau}{m}, \qquad \rho = 1/\sigma_o. \tag{3.13}$$

Here, n_e, e, and ρ are the electron density, electron charge, and bulk resistivity, respectively. τ is the mean carrier collision time. The primary difference between an insulator and a metal from this classical perspective is that the former has resonances at finite energy (i.e., typically in the infrared to optical) and the latter has a pole at zero frequency. This can be seen by considering the Lorentzian form and noting the expression reduces to the Drude dielectric function in the limit the restoring force tends to zero. This functional form is sufficient for the immediate purpose at hand.[1]

3.3 General Characterization of Dielectrics

Treatment of general dielectric materials, with loss included, requires the introduction of a complex dielectric function, $\varepsilon = \varepsilon' + i\varepsilon''$, where it is customary to indicate the real and imaginary components through the use of primed values. This convention primarily results from the notation, $\varepsilon_r = \varepsilon_r' + i\varepsilon_r''$, being reserved for the *relative* permittivity, $\varepsilon_r \equiv \varepsilon/\varepsilon_o$, where ε_o is the permittivity of free space.[2] The "+" sign reflects our choice for the harmonic dependence, $\exp(-i\omega t)$, of plane wave solutions (Palik 1998).

The Dielectric Function

Note that ε is casually referred to as the "dielectric constant"; however, in general, ε is a function of frequency, $\varepsilon(\omega)$. The only case in classical electrodynamics where ε is truly a constant is that of a vacuum. In this limit, the medium is lossless, and by the Kramers–Kronig optical sum rule the dielectric function is a constant. For brevity, we will often omit the explicit frequency dependence in the notation, but it is good to keep in mind that in general $\varepsilon = \varepsilon(\omega)$.

The relative permittivity is related to the (frequency-dependent) index of refraction, which is useful in describing the propagation and decay of the electric field within a lossy medium,

[1] Far-infrared photon energies typically lie below the scale where transitions between the valence and conduction bands have a significant probability of occurrence in metals (Ordal et al. 1985). In adopting a classical Drude–Lorentz functional form for conductive media, a weak energy dependence in the electron scattering time may be encountered in modeling observed spectra as this limit is approach. For a more refined physical treatment and discussion see Lucarini et al. (2005); Shiles et al. (1980) and references therein.

[2] ε_o is calculated from the defined speed of light, $c^2 = 1/(\varepsilon_o\mu_o)$, where $\mu_o \equiv 4\pi \times 10^{-7}$ H m^{-1} in MKS.

$$\tilde{n}(\omega) = n(\omega) + i\kappa(\omega).$$

Assuming for simplicity that the material is nonmagnetic, the relative permeability is $\mu_r \equiv \mu/\mu_o = 1$. We relate the real and imaginary parts of the complex index to those of the complex relative permittivity,

$$\begin{aligned} \tilde{n}^2 &= \varepsilon_r \\ (n + i\kappa)^2 &= \varepsilon_r' + i\varepsilon_r'' \\ n^2 - \kappa^2 + i2n\kappa &= \varepsilon_r' + i\varepsilon_r''. \end{aligned}$$

Matching the real and imaginary parts of the above equation yields

$$\begin{aligned} \varepsilon_r' &= n^2 - \kappa^2, \\ \varepsilon_r'' &= 2n\kappa. \end{aligned}$$

The dielectric loss tangent is the ratio of the conduction current relative to the displacement current in the media and can be expressed as

$$\tan\delta_e = \frac{\varepsilon_r''}{\varepsilon_r'} = \frac{2n\kappa}{n^2 - \kappa^2}.$$

Similarly, n and κ can be solved for and expressed in terms of the dielectric components ε_r' and ε_r'' as

$$n = \sqrt{\frac{1}{2}\left[\left(\varepsilon_r'^2 + \varepsilon_r''^2\right)^{1/2} + \varepsilon_r'\right]}, \tag{3.14}$$

$$\kappa = \sqrt{\frac{1}{2}\left[\left(\varepsilon_r'^2 + \varepsilon_r''^2\right)^{1/2} - \varepsilon_r'\right]}. \tag{3.15}$$

Because $\varepsilon(\omega)$ describes a physical quantity that has finite support in the frequency domain and the system must obey causality, the real and imaginary parts of the dielectric function are related by the Kramers–Kronig relations (Arfken 1985; Lucarini et al. 2005). For example, the real part can be written in terms of the Cauchy principal value ($\mathcal{P}$) of the integral of the imaginary part (Yariv & Yeh 2005; Jackson 1972),

$$[\varepsilon_r'(\omega_0) - 1] = +\frac{2}{\pi}\mathcal{P}\int_0^\infty \frac{\omega\varepsilon_r''(\omega)}{\omega^2 - \omega_0^2}d\omega. \tag{3.16}$$

Likewise, the inverse relationship is specified by a Hilbert transform,

$$\varepsilon_r''(\omega_0) = -\frac{2}{\pi}\mathcal{P}\int_0^\infty \frac{\omega_0[\varepsilon_r'(\omega) - 1]}{\omega^2 - \omega_0^2}d\omega, \tag{3.17}$$

where we have tacitly assumed the dielectric function is regular at zero frequency, as is the case for dielectrics.[3] This general result arises directly from enforcing causality

[3] For conductors, a simple pole at zero can be added. Notice the real and imaginary components of the dielectric function are even and odd functions of frequency, respectively.

between the polarization vector, **P**, in the media and electric field **E**, at all times *prior* to the measurement time. These integral equations directly connect the real and imaginary components of the dielectric function—they are not independent but two parts of the same entity. As an additional consequence, these relationships provide a path to the determination of self-consistent optical parameters, evaluation of uncertainties arising from finite observational spectral range, and the theoretical development of numerous sum rules and constraints (Lucarini et al. 2005).

For homogeneous and amorphous electromagnetic media the dielectric permittivity can be represented as a scalar quantity. More generally, in anisotropic media parameterization as a tensor related to the symmetry class of the media is necessary (Franta 2020). The dielectric permittivity tensor eigenvalues represent the material's permittivity along the principal axes. structure. If two eigenvalues are equal the material is designated "uniaxial," and is "biaxial" when all three differ in magnitude.

As an example, a birefringent crystal having its extraordinary axis aligned in the z-direction, the dielectric permittivity takes the form,

$$\tilde{\varepsilon}_\mathrm{r} = \begin{pmatrix} \varepsilon_\mathrm{r}^\mathrm{o} & 0 & 0 \\ 0 & \varepsilon_\mathrm{r}^\mathrm{o} & 0 \\ 0 & 0 & \varepsilon_\mathrm{r}^\mathrm{e} \end{pmatrix}.$$

This is a symmetric matrix, and the o and e labels indicate the "ordinary" and "extraordinary" axes of the material, respectively. The relationship between the applied electric field and the resultant electric displacement is specified by:

$$\mathbf{D} = \varepsilon_\mathrm{o}\tilde{\varepsilon}_\mathrm{r} \cdot \mathbf{E}. \tag{3.18}$$

For polarimetry, these quantities can be employed to describe biattenuance and birefringence, which can occur in either the linear or circular polarization bases. Linear birefringence and biattenuance are observed in crystals such as quartz, calcite, and sapphire, which are common materials for constructing optical elements such as wave plates. Circular birefringence and biattenuance are observed in optically active chiral media (e.g., dielectric structures and bulk media with intrinsic handedness) and Faraday active media (e.g., magnetic materials such as ferrites, interstellar magnetized plasma medium, etc.). Metamaterial techniques can be used to synthesize both conditions (Pisano et al. 2012; Wang et al. 2009) at frequencies as high as the terahertz region of the electromagnetic spectrum. Efforts to extend polarization into and beyond the visible is an ongoing metamaterial research thrust, largely limited by fabrication capability and achievable tolerances.

3.3.1 Skin Depth

A general wave has the following functional dependence along the direction of propagation (z):

$$\begin{aligned} \mathbf{E}(\omega,\mathrm{t}) \propto \exp(\mathrm{i}\gamma z) &= \exp(2\pi \mathrm{i} z\tilde{n}/\lambda_\mathrm{o}) \\ &= \exp(2\pi \mathrm{i} z(n + \mathrm{i}\kappa)/\lambda_\mathrm{o}) \\ &= \exp(2\pi \mathrm{i} zn/\lambda_\mathrm{o})\exp(-2\pi z\kappa) = \exp(\mathrm{i}\beta z)\exp(-z/\delta_s). \end{aligned}$$

Here, the complex propagation constant, γ, is introduced. Its real part is β, which is related to the guide wavelength, λ_o/n. The second factor in the resulting expression describes the attenuation of the field as it propagates through the material. The fields are confined within a layer characterized by *skin depth*, which is the distance the wave travels in the medium for which the field strength decreases by 1/e. This field penetration or extinction length scale is related to the imaginary part of the index of refraction in the metal by

$$\delta_s \equiv \frac{1}{2\pi\kappa}.$$

Note that for good conductors that are Ohmic ($\mathbf{J} = \sigma\mathbf{E}$), $\varepsilon'' \gg \varepsilon'$ (Pozar 2012; Leontovich 1948; Senior 1960),

$$\delta_s = \sqrt{\frac{2}{\omega\mu\sigma}}. \tag{3.19}$$

Here, σ is the bulk conductivity of the material, which can be related to the surface resistance, $R_s = 1/\sigma\delta$, of the bulk metal. The Hagen–Rubens relations (Hagen and Rubens 1903) provide a direct physical connection between the metal's optical response and its corresponding transmission line circuit representation, $A = 1 - R = 2R_s/\eta_o$, where A is the absorptance, R is the reflectance, and η_o is the impedance of free space (Chuss et al. 2012).

This can be particularly important in the context of thermal emission from the optical elements of an instrument at finite temperature. At normal incidence the emission from the horizontal and vertical polarizations are identical; however, at finite angle of incidence this degeneracy is broken as anticipated by the Fresnel equations, and the resulting thermal emission is polarized. In the context of polarimetry this physical mechanism can lead to instrumental polarization (Wollack et al. 1993, 1994) and polarized radiometric offsets in beam steering or chopping (Cortiglioni 1994). On the other hand, this phenomenon also provides opportunities for precision characterization of material properties (Xu et al. 1996; Renbarger et al. 1998) and instrument flux calibration (Bischoff et al. 2013; Barkats et al. 2005).

The skin depth for a good conductor (e.g., aluminum) at 30 GHz is $\delta \approx 0.5\ \mu$m. The field penetration scales as $\omega^{-1/2}$ through the far-infrared at which point, the electron-electron scattering time starts to roll off the Drude response (Lucyszyn 2004), and the onset of inter-band scattering is observed. Surface roughness and defects at the scale of the skin depth are known to influence reflective losses and by extension surface emission (Morgan 1949). Either of these effects can lead to polarized signatures associated with their degree of coherence and interaction with the incident fields with these features. A useful compilation of measured room temperature data for engineering metals at wavelengths longward of the far-infrared can be found in Ordal et al. (1983, 1985).

The above examples provide insight into how to employ surface impedance boundary conditions rather than modeling the entire physical volume of the media. This is an example of the reduction of complexity from a field theory problem (i.e., via perturbation of boundary conditions from mathematical physics) to a transmission line circuit representation.[4]

3.3.2 Surface Resistance

We can imagine a thin film having a finite conductivity, $\sigma = 1/\rho$, where ρ is the resistivity. The thickness (t) is constant, and we can imagine varying the height (h) and width (w) of the film's volume. Furthermore, we consider a static and constant electric field parallel to the sheet and oriented along the h dimension. This leads to a the presence of a current density, which flows through the conductive material. The total resistance presented by the sheet is

$$R_{\text{total}} = \frac{V}{I} = \frac{\int \mathbf{E} \cdot d\mathbf{l}}{\int \mathbf{J} \cdot d\mathbf{A}} = \frac{Eh}{\sigma Ewt} = \frac{h}{wt/\rho} = \frac{h}{w}\frac{\rho}{t} = \frac{h}{w}R_s$$

The resistance of the sheet can be characterized by the sheet resistance R_s. The other factor is just the aspect ratio h/w. In the layout of planar circuit, this is often quoted in terms of the number of squares along the direction of the electric field vector. Figure 3.1 shows two examples of different numbers of squares and thus different total resistances. In the context of electromagnetic plane wave propagation the unit cell is a square, $w = h$, and the characteristic impedance in the media is specified with units of "Ohms per square."

In going to finite frequency, a similar relationships holds; however, one needs to account for the penetration or diffusion of the electromagnetic fields into the metal

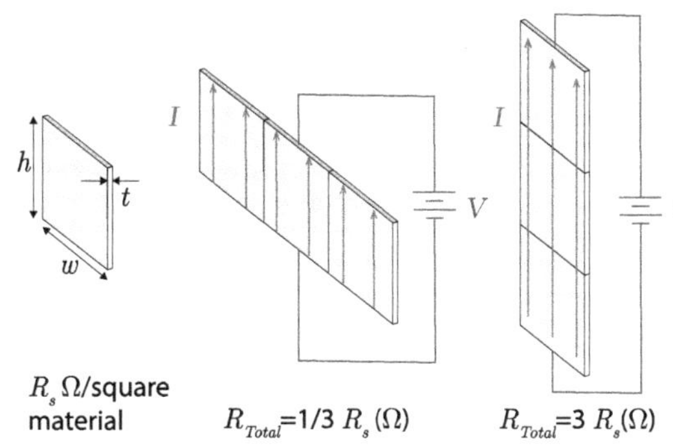

Figure 3.1. The concept of surface resistance at zero frequency is shown. A unit square is shown on the left and a 3 × 1 realization is shown in the middle and right panels. The resistance of the circuit depends on the orientation of the layer with respect to the electric field vector (or current flow as shown above).

[4] An example of this is the Wheeler incremental inductance rule (Wheeler 1942). This analysis strategy has more general applicability. In electromagnetic design, it is common for representative dimensions of elements to span an immense dynamic range. An understanding of the field structure within materials and devices can provide opportunities to significantly reduce computational resources required in modeling while sacrificing neither accuracy nor physical insight into the device's operation.

film, which is controlled by the skin depth. Assuming plane wave illumination of the film at normal incidence, one finds for a uniform square unit cell,

$$R_s = \frac{\rho}{\delta}\frac{1}{(1 - e^{-t/\delta})}, \tag{3.20}$$

where wave interference within the metal film has been ignored for simplicity (Jackson 1972). For films thicker than a few skin depths the second term in the denominator reduces to zero and the bulk properties of the material dominate, $R_s \rightarrow \rho/\delta$, and as the frequency tends to zero one recovers the static surface resistance for the film resistor, $R_s \rightarrow \rho/t$.

These strategies can be extended and employed in graphically estimating the characteristic impedance as well as the capacitance and inductance per unit length of transmission line structures by appropriately combining the number of squares in parallel and series (Kraus 1984; Von Hipple 1954). In the context of the numerical simulation of thin film devices, the use of a complex surface impedance representation for electromagnetic media (Kerr 1999) enables the efficient and accurate evaluation of the influence of kinetic inductance on the response. This is typically of interest in superconducting (U-Yen et al. 2018) and normal metal films (Chuss et al. 2012), where the field penetration scale relative the wavelength is finite. Other surface impedance and concepts which find utility include the perfect-E mirror (i.e., which excludes E-fields and represents a zero impedance or a "short") and the perfect-H mirror (i.e., an infinite impedance or an "open"), respectively, at a boundary. These symmetries find application in efficiency representing guiding, periodic, and antenna structures encountered in polarimetric systems.

More generally, the filling fraction of the unit cell array tiling and geometry of the film can be tailored to synthesize a complex surface impedance. The optical response is dependent upon the illumination polarization and angle of incidence relative to array. This class of structures can be used to absorb or modify the spectral and polarization content of a signal. Such planar structures are referred to as "polarization selective surfaces (PSS)," "frequency selective surfaces (FSS)" (Munk 2005), and "meta-materials" (Joannopoulos et al. 2008) in the literature. Similarly, in early microwave works, the terminology "artificial dielectrics" may be encountered for sub-wavelength volumetric meta-structures and devices (Collin 1990). Devices with structured elements that approach a wavelength in scale fall under the rubric of photonic crystals (Jovanovic et al. 2023), an emerging technique to realize multi-functional integrated instrument systems. Figure 3.2 shows an example of a unit cell representation of a PSS. In this simple example (for normal incidence), the polarizations are degenerate. At finite incidence angle, the system is polarization-dependent.

3.4 Wave Impedance

The electromagnetic constituent relations (i.e., the permittivity (ε), permeability (μ), and conductivity (σ)) relate the electric and magnetic fields to the media's displacement, magnetization, and current density vectors. These material parameters, the boundary conditions, and the detailed geometry allow solutions of the wave

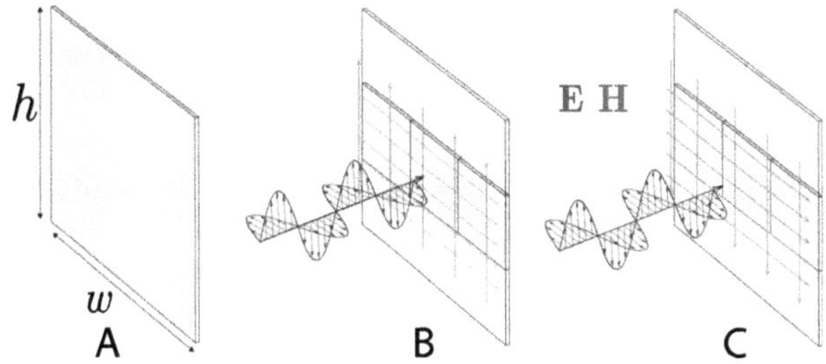

Figure 3.2. The surface impedance concept is shown for a (A) symmetric unit cell. In this case, light polarized in the vertical and horizontal direction will have the same surface impedance and thus the same reflectance properties. In (B) and (C), the unit cell shows an asymmetric geometry such that the vertical and horizontal impedances are different, and the surface is polarization-selective.

equation. The intrinsic impedance, η, is derived from Maxwell's equations and physically corresponds to the ratio of the magnitude of the electric field relative to that of the magnetic field. For a homogeneous media and plane wave propagation, $\eta(\omega) \equiv |\mathbf{E}|/|\mathbf{H}| = \sqrt{\mu(\omega)/\varepsilon(\omega)}$, is commonly referred to as the *wave impedance*.

In free space, the wave impedance's magnitude can be expressed, $\eta_{\text{o}} = \sqrt{\mu_{\text{o}}/\varepsilon_{\text{o}}} = 377\ \Omega$ per square, where the permeability and permittivity of free space are, $\mu_{\text{o}} \equiv 4\pi \times 10^{-7}$ H m^{-1} and $\varepsilon_{\text{o}} = 8.854 \times 10^{-12}$ F m^{-1}, respectively. Similar to the speed of light, $c = \sqrt{1/\varepsilon_{\text{o}}\mu_{\text{o}}}$, the impedance of free space is a real-valued parameter that is independent of frequency and *non-dispersive*. Hence, it is lossless. More generally, η is a complex, frequency-dependent quantity that defines the coupling between E and H in the medium. The refractive index, $\tilde{n}$, specifies the relative reduction in the propagation velocity of light in the medium and losses arising from the presence of matter.

The ratio of the wave impedances at boundaries, or the impedance contrast, is a measure of the electromagnetic confinement and/or discontinuity experienced by a wave at an interface between dissimilar electromagnetic media. At interfaces between differing homogeneous media, reflection and transmission of the wave is set by the wave impedance magnitude and phase encountered by the signal. It is notable that the permeability, $\mu = \eta n$, and permittivity, $\varepsilon = n/\eta$, are fundamentally linked by the underlying structure and form of Maxwell's equations. One can either focus on the constitutive relations in a given medium, $(\varepsilon(\omega), \mu(\omega))$, or wave propagation in the medium, $(n(\omega), \eta(\omega))$, to parameterize the electromagnetic response. However, in using both, the constraints set by these relationships must be respected in extraction of self-consistent transmission line and circuit model parameters (e.g., see discussion in Cataldo et al. (2015)).

3.5 Reflections at an Interface: The Fresnel Equations

Electromagnetic wave propagation across a boundary between different materials can be treated using Maxwell's equations. This problem has many important applications as it underlies fundamental optics principles (such as the laws of reflection and refraction) because in general both a transmitted and reflected wave are required in order to satisfy the boundary conditions presented. The details of the direction and magnitude of the reflected and transmitted waves are material-dependent, but most relevant materials are well-modeled using the properties of a general dielectric. For non-normal incidence, polarization components perpendicular and parallel to the plane of incidence experience different field configurations and lead to different fractions of their power being reflected and transmitted. Thus, oblique reflections in an optical system can lead to the introduction of polarization, even if the incoming radiation is unpolarized. This is an example of asymmetry leading to polarization. This basic principle extends to astrophysical problems. For example, the reflection of starlight via the atmosphere of an exoplanet is predicted to produce polarization.

The key equations that model the relative amplitudes of the reflected and transmitted fields as a function of material properties and geometry (incidence angle) are known as the *Fresnel equations*. The Fresnel equations are specifically useful for handling reflection at a boundary between dielectrics, which is often a key consideration in the design and characterization of polarimetry instrumentation. Similarly, the formalism provide a means to evaluate polarized emission from the surface of reflective optics, modulators, and flux calibration sources.

One of the most common sources of instrumental polarization arises from reflections within the optical components of the telescope. Such a phenomenon might be obvious in the case of an oblique reflection from, for example, a flat tertiary mirror; however, any reflection may be a source of (unwanted) instrumental polarization, specifically when the optical symmetry is broken. Glint and ghost images, which result from imperfect illumination and unwanted reflections in the optical system can both be polarized as a result of such processes.

Because of the utility of the Fresnel equations to instrumentation, we include a derivation of the Fresnel equations here. Such an argument can be found in many other text on electromagnetics and electrical engineering (Pozar 2012; Jackson 1972; Hecht 2017). In these derivations, we assume that the dielectric are lossless, and thus the dielectric functions and consequently the index of refraction are real. This can be easily extended to lossy materials via the use of a complex dielectric function.

Figure 3.3 shows the problem at hand. A plane wave is incident from the left on the dielectric boundary between a region having permittivity and permeability ε_1 and μ_1, respectively, and a region characterized by ε_2, μ_2. The plane wave's **k**-vector makes an angle θ_I with the normal to the boundary. There is a reflected wave making an angle of θ_R with the normal, and a transmitted wave at an angle of θ_T with the normal. The reason for the reflected wave is exactly the same as for an unmatched transmission line. The impedance of each of the two regions ($\sqrt{\mu/\varepsilon}$) is different, so a reflected electric field is necessary to match the boundary conditions. This will arise in the course of the analysis.

As a reminder of the comment in the previous chapter, it is often useful to consider the **E** and **H** fields. This enables a manifest analogy to circuit theory since the ratio of E to H has units of Ohms and can be considered as the impedance of the

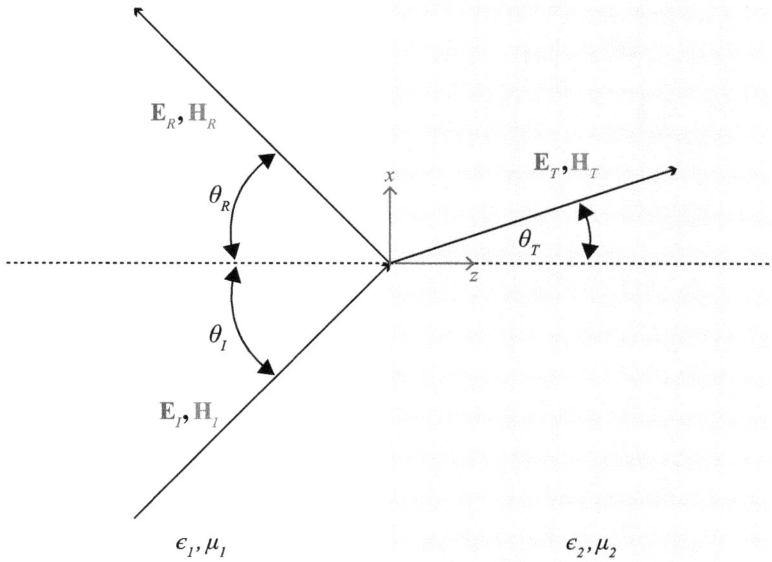

Figure 3.3. The geometry and notation for the fields associated with a radiation crossing a boundary.

radiative mode in question. In addition, the Poynting vector can be written as $\mathbf{S} = \mathbf{E} \times \mathbf{H}$ without the need for including the permeability of the medium explicitly. Keeping in mind that using $\mathbf{D} = \varepsilon\mathbf{E}$ and $\mathbf{B} = \mu\mathbf{H}$, we can show the continuity of the appropriate fields. Maxwell's equations in integral form (Jackson 1972) provide boundary conditions for the fields at the interface. The arguments for the boundary conditions for the parallel and normal fields are illustrated in Figure 3.4.

For the **D** and **B** fields, we utilize Gauss' Law for magnetism and create a cylindrical Gaussian surface across the boundary. This is shown in the top row of Figure 3.4. Since the total integral of the scalar product of the field and the normal over the entire surface must be zero in each case,[5] we can shrink the height of the cylinder to zero and note that the perpendicular component of each field must be continuous across the interface.

For the **E** and **H** fields, we construct a rectangular path in a plane perpendicular to the interface (see the bottom row of Figure 3.4). In this case, the line integral for each of these fields must be zero. We shrink the perpendicular side of the path to zero, and find that the tangential components of these fields must be continuous across the boundary.

To summarize, we can write the boundary conditions at the interface as

$$\varepsilon_1 E_1^{\perp} = \varepsilon_2 E_2^{\perp}, \tag{3.21}$$

[5] For the **D** field, there are no free charges in either dielectric; for the **B** field there are no monopoles.

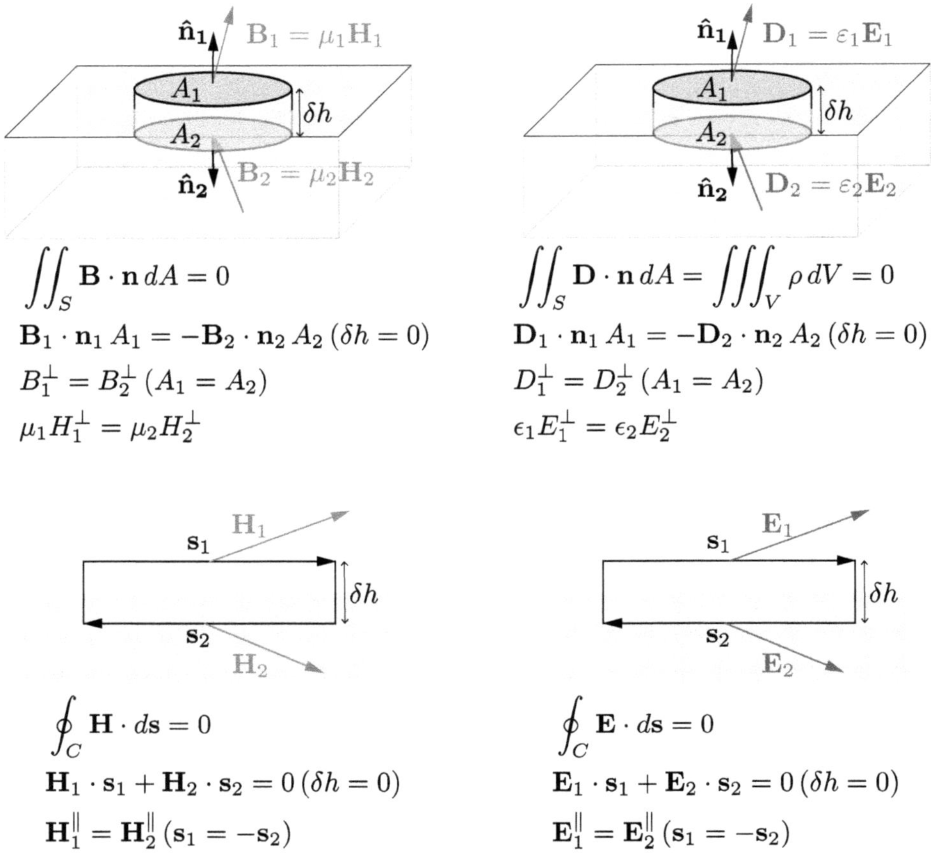

Figure 3.4. Gaussian surfaces and path integrals constructed at the interface between two dielectrics demonstrate the constraints on the **E** and **H** fields across the boundary.

$$\mu_1 H_1^{\perp} = \mu_2 H_2^{\perp}, \tag{3.22}$$

$$\mathbf{E}_1^{\parallel} = \mathbf{E}_2^{\parallel}, \tag{3.23}$$

$$\mathbf{H}_1^{\parallel} = \mathbf{H}_2^{\parallel}. \tag{3.24}$$

We assume general solutions to the wave equations as shown in Figure 3.3 for incident, reflected and transmitted waves orphaned equations,

$$\mathbf{E}_I(\mathbf{r}, t) = \mathbf{E}_{0\mathrm{I}} e^{i(\mathbf{k}_\mathrm{I} \cdot \mathbf{r} - \omega_\mathrm{I} t)}, \tag{3.25}$$

$$\mathbf{E}_R(\mathbf{r}, t) = \mathbf{E}_{0\mathrm{R}} e^{i(\mathbf{k}_\mathrm{R} \cdot \mathbf{r} - \omega_\mathrm{R} t + \varepsilon_\mathrm{R})}, \tag{3.26}$$

$$\mathbf{E}_T(\mathbf{r}, t) = \mathbf{E}_{0\mathrm{T}} e^{i(\mathbf{k}_\mathrm{T} \cdot \mathbf{r} - \omega_\mathrm{T} t + \varepsilon_\mathrm{T})}. \tag{3.27}$$

Here, ε_R and ε_T are relative phases of the reflected and transmitted wave relative to the incident wave. The corresponding **H**-fields are

$$\mathbf{H}_I(\mathbf{r}, t) = \frac{1}{\eta_I}(\hat{\mathbf{k}}_I \times \mathbf{E}_I), \tag{3.28}$$

$$\mathbf{H}_R(\mathbf{r}, t) = \frac{1}{\eta_R}(\hat{\mathbf{k}}_R \times \mathbf{E}_R), \tag{3.29}$$

$$\mathbf{H}_T(\mathbf{r}, t) = \frac{1}{\eta_T}(\hat{\mathbf{k}}_T \times \mathbf{E}_T). \tag{3.30}$$

Here, $\eta_I = \eta_R = \sqrt{\mu_1/\varepsilon_1}$ is the wave impedance of material 1 and $\eta_T = \sqrt{\mu_2/\varepsilon_2}$ is the wave impedance of material 2.

At the boundary, the arguments for the exponential factors must all be equal. This leads to three consequences. First, because the temporal part of the argument (ωt) must be equal for the three waves, frequency conversion does not occur.[6] That is, the fields for each of the waves must oscillate at the same frequency or the boundary conditions will not match. The equality of the spatial part of the exponentials for the three waves leads to several important consequences. First, the three k-vectors must lie in a plane. To see this, we can use the fact that the parallel component of the electric field is continuous across the boundary,

$$\hat{\mathbf{n}} \times \mathbf{E}_I + \hat{\mathbf{n}} \times \mathbf{E}_R = \hat{\mathbf{n}} \times \mathbf{E}_T \tag{3.31}$$

$$\hat{\mathbf{n}} \times \mathbf{E}_{0I} e^{i(\mathbf{k}_I \cdot \mathbf{r} - \omega_I t)} + \hat{\mathbf{n}} \times \mathbf{E}_{0R} e^{i(\mathbf{k}_R \cdot \mathbf{r} - \omega_R t + \varepsilon_R)} = \hat{\mathbf{n}} \times \mathbf{E}_{0T} e^{i(\mathbf{k}_T \cdot \mathbf{r} - \omega_T t + \varepsilon_T)}. \tag{3.32}$$

At the surface of the interface, these functions must have the same dependence on r and t. This means the arguments of the exponentials all need to be equal,

$$(\mathbf{k}_I \cdot \mathbf{r} - \omega_I t) = (\mathbf{k}_R \cdot \mathbf{r} - \omega_R t + \varepsilon_R) = (\mathbf{k}_T \cdot \mathbf{r} - \omega_T t + \varepsilon_T). \tag{3.33}$$

The time components must be the same, so this implies $\omega \equiv \omega_I = \omega_R = \omega_T$. The remaining parts of the argument must also be equal,

$$(\mathbf{k}_I \cdot \mathbf{r}) = (\mathbf{k}_R \cdot \mathbf{r} + \varepsilon_R) = (\mathbf{k}_T \cdot \mathbf{r} + \varepsilon_T).$$

The phase factors here are indicative of the freedom to choose the origin of the spatial coordinate system. We can set these equal in pairs,

$$(\mathbf{k}_I \cdot \mathbf{r}) = (\mathbf{k}_R \cdot \mathbf{r} + \varepsilon_R) \tag{3.34}$$

$$(\mathbf{k}_I - \mathbf{k}_R) \cdot \mathbf{r} = \varepsilon_R. \tag{3.35}$$

Here, the **r** terminates on the interface. This is the equation of a plane. The meaning of this equation is most obvious if we take the origin of the coordinate system to be

[6] Here, for simplicity, we assume that the media are linear, the boundary is stationary, and the incident radiation is a monochromatic plane wave. This can be readily generalized to treat a spectrum of modes at an interface. This approach goes under the terminology "mode matching" in the literature for guided waves.

in the plane of the interface. In this case, $\varepsilon_R = 0$. Also noting that the magnitudes of the two k-vectors are the same (since they are in the same medium), $\mathbf{k}_I - \mathbf{k}_R$ is perpendicular to the interface. This leads to

$$\begin{aligned} \hat{\mathbf{n}} \times (\mathbf{k}_I - \mathbf{k}_R) &= 0 \\ k_I \sin\theta_I &= k_R \sin\theta_R. \end{aligned}$$

Here, θ_I and θ_R are the incident and reflected angles, respectively. This implies the law of reflection, $\theta_I = \theta_R$. A similar argument implies that

$$\begin{aligned} \hat{\mathbf{n}} \times (\mathbf{k}_I - \mathbf{k}_R) &= 0 \\ k_I \sin\theta_I &= k_T \sin\theta_T \end{aligned}$$

In this case, the k's are proportional to the indices of refraction, and we recover Snell's law,

$$\frac{\sin\theta_T}{\sin\theta_I} = \frac{k_1}{k_2} = \frac{n_1}{n_2}.$$

3.5.1 Polarization Perpendicular to the Plane of Incidence

Figure 3.5 shows the case for polarization perpendicular to the plane of incidence. This is often called the s-wave[7] or the "TE (transverse electric)" mode because the E-fields are entirely perpendicular (transverse) to the plane of incidence (i.e., parallel to the interface).

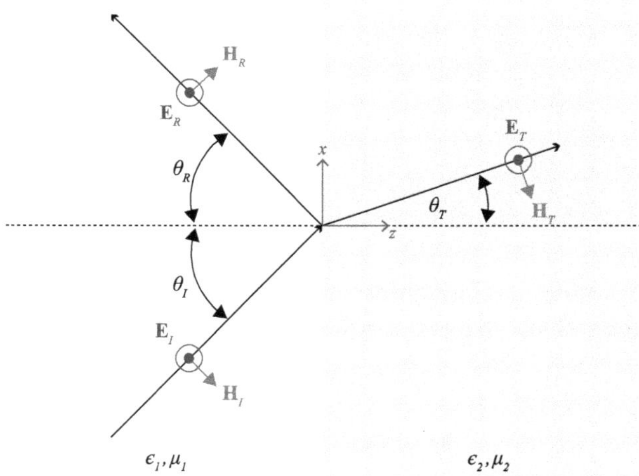

Figure 3.5. The field configuration is shown for the case where the electric field is parallel to the boundary and perpendicular to the plane of incidence (TE Mode).

[7] The terminology "s-polarization" comes from the German *senkrecht*, meaning "perpendicular," indicating that the E fields are perpendicular to the plane of incidence.

We take the z-direction to be perpendicular to the interface, with x in the upward direction. The fields in this case are

$$B_\perp\text{:}\ \frac{1}{\eta_1\mu_1}(E_{0\text{I}}\sin\theta_\text{I} + E_{0\text{R}}\sin\theta_\text{R}) = \frac{1}{\eta_2\mu_2}E_{0\text{T}}\sin\theta_\text{T} \tag{3.36}$$

$$E_\parallel\text{:}\ E_{0\text{I}} + E_{0\text{R}} = E_{0\text{T}} \tag{3.37}$$

$$H_\parallel\text{:}\ \frac{1}{\eta_1}(-E_{0\text{I}}\cos\theta_\text{I} + E_{0\text{R}}\cos\theta_\text{R}) = -\frac{1}{\eta_2}E_{0\text{T}}\cos\theta_\text{T}. \tag{3.38}$$

Defining the reflection and transmission coefficients as $r \equiv E_{0\text{R}}/E_{0\text{I}}$ and $t \equiv E_{0\text{T}}/E_{0\text{I}}$, respectively (also recall $\theta_\text{I} = \theta_\text{R}$), gives

$$B_\perp\text{:}\ \frac{1}{\eta_1\mu_1}(\sin\theta_\text{I} + r\sin\theta_\text{I}) = \frac{1}{\eta_2\mu_2}t\sin\theta_\text{T} \tag{3.39}$$

$$E_\parallel\text{:}\ 1 + r = t \tag{3.40}$$

$$H_\parallel\text{:}\ \frac{1}{\eta_1}(-\cos\theta_\text{I} + r\cos\theta_\text{I}) = -\frac{1}{\eta_2}t\cos\theta_\text{T}. \tag{3.41}$$

We can use the second two equations to solve for r and t.

$$r = \frac{\eta_2\cos\theta_\text{I} - \eta_1\cos\theta_\text{T}}{\eta_2\cos\theta_\text{I} + \eta_1\cos\theta_\text{T}}, \tag{3.42}$$

$$t = \frac{2\eta_2\cos\theta_\text{I}}{\eta_2\cos\theta_\text{I} + \eta_1\cos\theta_\text{T}}. \tag{3.43}$$

For nonmagnetic materials, the wave impedance in each material can be expressed in terms of the index of refraction of each $\eta_1 = \eta_\text{o}/n_1$, $\eta_2 = \eta_\text{o}/n_2$. This provides alternate expressions,

$$r_\perp = \frac{n_1\cos\theta_\text{I} - n_2\cos\theta_\text{T}}{n_1\cos\theta_\text{I} + n_2\cos\theta_\text{T}} \tag{3.44}$$

$$t_\perp = \frac{2n_1\cos\theta_\text{I}}{n_1\cos\theta_\text{I} + n_2\cos\theta_\text{T}}. \tag{3.45}$$

3.5.2 Polarization Parallel to the Plane of Incidence

A similar analysis can be done for the case in which the electric field is in the plane of incidence as shown in Figure 3.6. This is often called the p-wave[8] or the "TM

[8] The terminology "p-polarization" comes from the German word "*parallel*" indicating that the E fields are parallel to the plane of incidence.

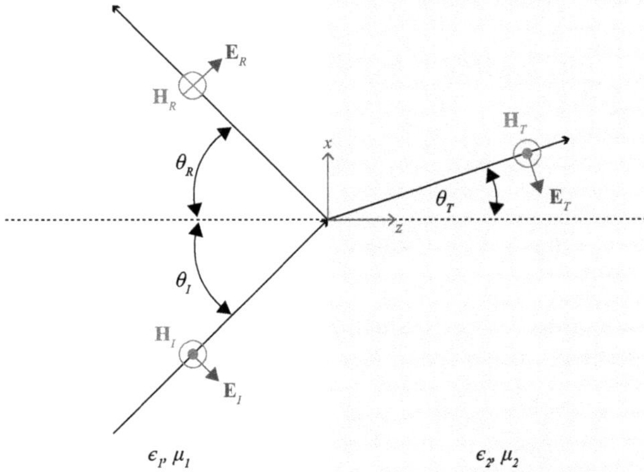

Figure 3.6. The field configuration is shown for the case where the electric field is parallel to the plane of incident (TM Mode).

(transverse magnetic)" mode because the H-fields are parallel to the interface (E-fields are parallel to the plane of incidence).

In this case, the boundary conditions can be expressed as

$$D_\perp\text{: } \varepsilon_1(E_{0\mathrm{I}} \sin\theta_\mathrm{I} + E_{0\mathrm{R}} \sin\theta_\mathrm{R}) = \varepsilon_2 E_{0\mathrm{T}} \sin\theta_\mathrm{T} \tag{3.46}$$

$$E_\parallel\text{: } E_{0\mathrm{I}} \cos\theta_\mathrm{I} - E_{0\mathrm{R}} \cos\theta_\mathrm{R} = E_{0\mathrm{T}} \cos\theta_\mathrm{T} \tag{3.47}$$

$$H_\parallel\text{: } H_{0\mathrm{I}} + H_{0\mathrm{R}} = H_{0\mathrm{T}}. \tag{3.48}$$

Again, defining the reflection coefficients the same way as above,

$$D_\perp\text{: } \varepsilon_1(\sin\theta_\mathrm{I} + r\sin\theta_\mathrm{I}) = \varepsilon_2 t \sin\theta_T \tag{3.49}$$

$$E_\parallel\text{: } \cos\theta_\mathrm{I} - r\cos\theta_\mathrm{I} = t\cos\theta_\mathrm{T} \tag{3.50}$$

$$H_\parallel\text{: } \frac{1}{\eta_1} + r\frac{1}{\eta_1} = t\frac{1}{\eta_2}. \tag{3.51}$$

Defining the reflection and transmission coefficients as $r = E_{0\mathrm{R}}/E_{0\mathrm{I}}$ and $t \equiv E_{0\mathrm{T}}/E_{0\mathrm{I}}$, respectively, gives

$$r = \frac{\eta_2 \cos\theta_\mathrm{T} - \eta_1 \cos\theta_\mathrm{I}}{\eta_2 \cos\theta_\mathrm{T} + \eta_1 \cos\theta_\mathrm{I}}, \tag{3.52}$$

$$t = \frac{2\eta_2 \cos\theta_\mathrm{I}}{\eta_2 \cos\theta_\mathrm{T} + \eta_1 \cos\theta_\mathrm{I}}. \tag{3.53}$$

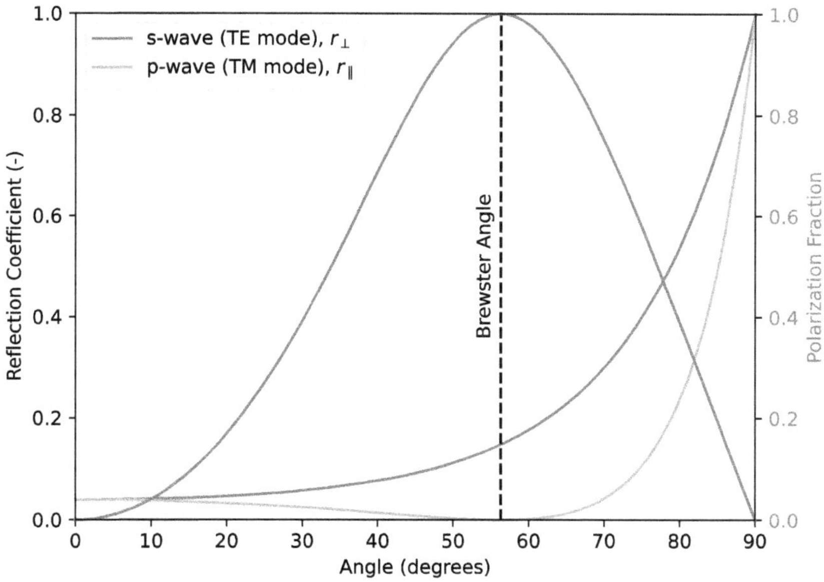

Figure 3.7. For an index $n = 1.5$, the reflection coefficients for both the s(TE) and p(TM) waves are shown. The fractional polarization for the reflected wave is plotted, peaking at the Brewster angle.

For nonmagnetic materials, the wave impedance in each material can be expressed in terms of the index of refraction of each $\eta_1 = \eta_o/n_1$, $\eta_2 = \eta_o/n_2$. This provides alternate expressions,

$$r_\parallel = \frac{-n_1 \cos\theta_T + n_2 \cos\theta_I}{n_1 \cos\theta_T + n_2 \cos\theta_I} \tag{3.54}$$

$$t_\parallel = \frac{2n_1 \cos\theta_I}{n_1 \cos\theta_T + n_2 \cos\theta_I}. \tag{3.55}$$

In each of these cases, θ_T can be calculated from the incidence angle via Snell's law.

When the incidence angle $\theta_I = 0$, the reflection coefficients are equal (as expected due to symmetry),

$$r_\parallel = r_\perp = \frac{n_2 - n_1}{n_2 + n_1}. \tag{3.56}$$

If $n_2 > n_1$, the coefficient is positive, meaning that there is zero phase shift upon reflection. If $n_T < n_I$, there is a 180° phase shift upon reflection.

Figure 3.7 shows the reflection coefficients for both TM and TE waves along with the fractional polarization of the reflected wave as a function of incident angle. In this case, the index of refraction is 1.5.

3.5.3 Brewster Angle

In the case where $n_2 > n_1$, $r_\parallel$ can go to zero at a particular angle. This angle is known as the *Brewster angle*. This happens when $n_1 \cos\theta_T = n_2 \cos\theta_I$. We can also keep in

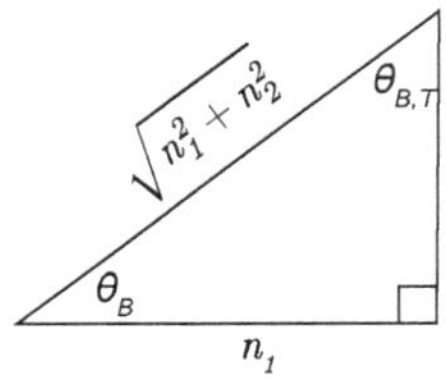

Figure 3.8. Geometric view of the Brewster angle.

mind Snell's law, $n_1 \sin\theta_I = n_2 \sin\theta_T$. If we simply solve these equations, we end up finding the solution $n_1 = n_2$. This is of course trivially the case when the media are the identical, there is no reflection because there is not an interface. To find the more interesting solution, imagine the situation shown in Figure 3.8 for which the incidence angle θ_B is complementary to the refracted angle $\theta_{B,T}$. In this case,

$$\cos\theta_B = \frac{n_1}{\sqrt{n_1^2 + n_2^2}} = \frac{n_1}{n_2}\cos\theta_{B,T}.$$

This situation satisfies the above condition. It also satisfies Snell's law because

$$\sin\theta_B = \frac{n_2}{\sqrt{n_1^2 + n_2^2}} = \frac{n_2}{n_1}\sin\theta_{B,T}. \tag{3.57}$$

Thus, by the geometry of the diagram, we can see that

$$\tan\theta_B = \frac{n_2}{n_1}. \tag{3.58}$$

When viewed at the Brewster angle, light becomes 100% polarized because the polarization that is parallel to the plane of incidence is completely suppressed. The Brewster angle is shown in Figure 3.7. The Brewster angle occurs at the angle for which the incident and transmitted angles are complementary, i.e., $\theta_B + \theta_{B,T} = \pi/2$.

3.6 Modeling Multilayers of Dielectrics

For various applications, multiple dielectric layers are stacked in an optical system. For optical components operating in transmission such as lenses and wave plates, anti-reflective coatings are often required to ensure effective transmission of the optical system. In this case, the desired effect is dependent upon the interference of multiple reflections to cancel the reflected wave via carefully-designed destructive interference. In a birefringent device such as a wave plate, the transmission will be polarization sensitive. There are other cases for which the desired effect of stacked dielectrics does not rely on interference of multiple reflections but simply utilize the propagation of the incident radiation. Nevertheless, the interference of multiple reflections in the structure usually must be understood and quantified to understand the systematics of the device. An example of this in polarimetry is the Pancharatnam wave plate (Pancharatnam 1955) architecture for which multiple wave plates are stacked at specified relative rotations to increase the bandwidth of the device. We will examine this concept in more detail in Chapter 4.

For such cases, the boundary conditions can be calculated at each interface, leading to an analytical expression for transmission and reflection. In practice, such expressions become cumbersome after only a few layers. Matrix techniques exist that allow the boundary conditions to be codified and organized for ease of computation and simulation.

Multilayer dielectrics can be analyzed analogously to transmission (ABCD) matrices (Pozar 2012). However, instead of a 2×2 matrix system that codifies the boundary conditions for single mode transmission lines in terms of the current and voltage, the quasioptical system requires a 4×4 system of matrices, which encodes the response to the **E** and **H** fields for each polarization. This formalism naturally captures the same physics as consideration of multiple reflections through the system.

This type of technique has been applied in the analysis of circular polarizers consisting of stacked birefringent layers (Adachi & Kennaugh 1960). More recently, a similar technique has been applied (with the addition of consideration of loss) by Savini et al. (2006) to a stack of dielectrics used as a broadband half wave plate. Essinger-Hileman (2013) also develops a transfer matrix approach to multilayered dielectrics. We refer the reader to these papers and references therein for detailed examples. Here we briefly outline the idea. The fields (using the notation in Savini et al. 2006) are represented by a column vector

$$\boldsymbol{\Psi}(z) = \begin{pmatrix} E_x(z) \\ H_y(z) \\ E_y(z) \\ -H_x(z) \end{pmatrix}. \tag{3.59}$$

The operation of a dielectric layer can be represented by a 4×4 propagation matrix, $\hat{\mathbf{P}}$, that relates the fields on one side of the layer to the other. This is typically dependent on the anisotropic birefringent and biattenuant properties of the material. The matrix $\hat{\mathbf{P}}$ can be expressed as a product of individual matrices representing the birefringence and general rotation matrices, $\mathbf{R}$,

$$\boldsymbol{\Psi}(z_1) = \hat{\mathbf{P}}\boldsymbol{\Psi}(z_1) = \prod_i \left(\mathbf{R}(\theta_i)\hat{\mathbf{Q}}_i\mathbf{R}^T(\theta_i)\right)\boldsymbol{\Psi}(z_2). \tag{3.60}$$

Here we are assuming propagation in the z−direction from z_1 to z_2 such that the rotation matrices $\mathbf{R}(\theta_i)$ represent rotations about the z-direction. The angle θ_i is that of the principal axis for layer i. The matrices $\hat{\mathbf{Q}}_i$ are block diagonal and encode the information regarding the loss and the propagation properties (via the dielectric matrix) of the extraordinary and ordinary axes of the birefringent material. Note that this technique assumes that one is in the *paraxial limit*, in which the fields for all relevant optical rays are perpendicular to the propagation system. In optical systems, this is equivalent to stating that the focal number, $f/$, is high.

3.7 Scattering

The physics of scattering is multi-faceted, and the mathematical tools that are convenient to use depend upon the quantity $x \equiv \frac{2\pi a}{\lambda}$. Here, a is the radius of a spherical particle (or some characteristic size of the particle for the non-spherical case), and λ is the wavelength of the light. In the context of this work, scattering of electromagnetic waves is a class of physical interactions that often leads to the introduction or modification of polarization. We touch on a few of these processes here. There are three main regimes that are usually considered. See van de Hulst (1957) for a more complete description and discussion.

- If $x \gg 1$, the scattering is *geometric*. Multiple coherent reflections usually can be ignored, and the problem can be considered using the Fresnel equations for a plane wave incident at the tangent of the particles surface.
- For the case where $x \sim 1$, the situation is more complex, since the details of the geometry and index of refraction can cause resonances inside the particle to be excited. In this regime, scattering blends into the concept of diffraction. Mathematically, this case can be treated using the formalism known as *Mie Scattering* (Bohren & Huffman 1983), which relies on the matching of spherical waves, both inside and outside of the particle to an incident plane wave using Maxwell's equations to set the conditions at the boundary. This limit also corresponds to systems for which the size of optical elements is similar to the wavelength. For instruments in which few modes are present, polarization effects must be accounted for because diffraction patterns (which include the beam pattern from antennas and feedhorns) are polarized as a function of angular location within the beam. This will be treated in more detail in the discussion of cross-polarization in the next chapter.
- For the case where $x \ll 1$, the details of the particle shape are less important, and most small particles can be effectively treated as radiating dipoles. This formalism is known as *Rayleigh scattering* and is the phenomenon responsible for the blue color of the sky.
- If the particles are electrons, the process is called *Thomson scattering* in the non-relativistic limit. If the electrons are relativistic, it is known as *Compton scattering*.

All of these processes can lead to polarization, since they involve an anisotropy between the emitter, scatterer, and observer. For example, will see in later chapters that Thomson/Compton scattering is the physical mechanism by which the cosmic microwave background becomes polarized.

3.7.1 Rayleigh Scattering

Rayleigh scattering is a polarization-sensitive process that applies in a wide variety of physical systems. In this case, the scattering particle has an induced dipole moment, $\mathbf{p}$ that is related to the incident electric field, $\mathbf{E}_0$ according to

$$\mathbf{p} = \alpha \mathbf{E}_0. \tag{3.61}$$

Generally, α is a tensor, but we consider here the case where the material is isotropic, so α can be reduced to a scalar quantity. From electromagnetic theory, the magnitude of the electric field produced at a large distance $r \gg \lambda$ away from a dipole is

$$E = \frac{k^2 p \sin\gamma}{r} \mathrm{e}^{-\mathrm{i}kr}, \tag{3.62}$$

where $k = 2\pi/\lambda$ and γ is the angle measured from the direction of the dipole moment. This "donut-shaped" radiation pattern is isotropic in the plane perpendicular to the dipole moment and goes to zero in the direction parallel to the dipole. For an incident plane wave in the z-direction, dipoles get introduced in both the x- and y-directions. This leads to two overlapping "doughnut" patterns rotated by 90° around the direction of propagation. The vertical polarization is then isotropic in the scattering plane. The horizontal polarization will be maximum in the forward and reverse directions, and will go to zero at a scattering angle of 90°. The vertical and horizontal intensities are then

$$I_{\mathrm{V}} = \frac{k^4 p^2}{r^2}, \tag{3.63}$$

$$I_{\mathrm{H}} = \frac{k^4 p^2 \cos^2\theta}{r^2}, \tag{3.64}$$

and the polarization fraction is

$$Q \equiv \frac{I_{\mathrm{V}} - I_{\mathrm{H}}}{I_{\mathrm{V}} + I_{\mathrm{H}}} = \frac{1 - \cos^2\theta}{1 + \cos^2\theta}. \tag{3.65}$$

Figure 3.9 shows these equations graphically.

3.7.2 Thomson Scattering

Thomson scattering refers to the scattering of radiation by an electron in the low energy limit. This corresponds to the case of $h\nu \ll m_{\mathrm{e}}c^2$. Here, ν is the frequency of the incident radiation, and m_{e} is the mass of the electron. In the limit where this condition is not met, the process must be treated as a photon-electron collision. This formalism is Compton scattering. The rest energy of an electron is 0.511 MeV c^{-2}, which corresponds to a frequency of 1.2×10^{11} THz, which is significantly beyond the spectral range covered in this book. Therefore, Thomson scattering adequately describes any EM-electron interactions that are relevant in this context.

In the dipole approximation, the differential power emitted by a dipole with a nonzero second time derivative is

$$\frac{dP}{d\Omega} = \frac{\langle \ddot{d} \rangle}{(4\pi)^2 \varepsilon_0 c^3} \sin^2\Theta. \tag{3.66}$$

Here, the angle brackets denote the time average.

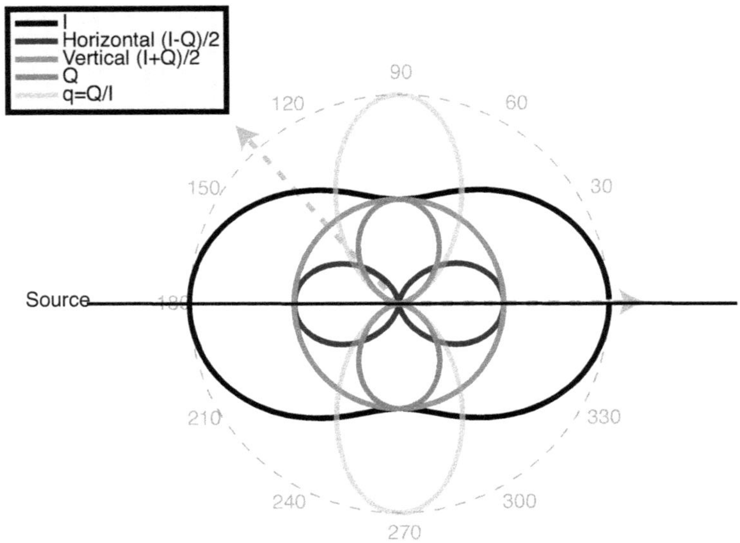

Figure 3.9. In Rayleigh scattering, the scattering particles are much smaller than the wavelength of light. In this case, the particle can be considered a point dipole. The polarization observed depends upon the angle between the direction of the sourced photons and the observer. In this case, the forward direction is defined to be 0° and the angle increases counterclockwise.

For Thomson scattering, the dipole moment of the electron is given by the product of its charge and the displacement that is induced by an incident plane wave. The force on the electron is just the time-varying electric field times its charge,

$$\mathbf{F} = m\ddot{\mathbf{r}} = e\mathbf{E}.$$

We take the incident field to be

$$\mathbf{E} = \hat{e}E_0 \sin \omega t, \tag{3.67}$$

where we have assumed a plane wave of frequency, ω with polarization in the $\hat{\varepsilon}$ direction.

The second derivative of the dipole is

$$\ddot{d} = e\ddot{r} = \frac{e^2 E_0}{m} \sin \omega t. \tag{3.68}$$

We next take the time average of the $\sin^2 \omega t$, which is $\frac{1}{2}$. The fractional cross-section is

$$\frac{dP}{d\Omega} = \frac{cE_0^2}{8\pi}\frac{d\sigma_\mathrm{T}}{d\Omega} = \frac{cE_0^2}{8\pi}\left[\frac{e^2}{4\pi\varepsilon_0 mc^2}\right]^2 \sin^2 \Theta. \tag{3.69}$$

It is customary to define the quantity r_e (the classical radius of the electron) as the length scale for which the electric energy is equal to the rest energy of the electron

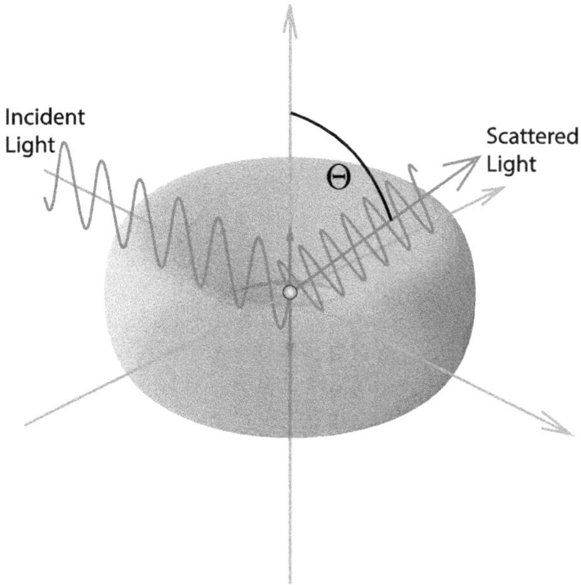

Figure 3.10. The geometry for Thomson scattering is shown. The yellow feature is the differential Thomson cross-section ($\frac{d\sigma_T}{d\Omega}$).

$$r_e \equiv \frac{e^2}{4\pi\varepsilon_o mc^2},$$

and the differential cross-section in terms of r_e is

$$\frac{d\sigma_T}{d\Omega} = r_e^2 \sin^2\Theta,$$

where Θ is the angle between the oscillation direction and the direction of the observer. The total integrated cross-section is

$$\sigma_T = \frac{8\pi}{3} r_e^2. \tag{3.70}$$

The angular distribution and geometry of Thomson scattering is shown in Figure 3.10.

3.8 Synchrotron Radiation

Synchrotron radiation is a complex topic that has been treated in the literature extensively. For a good textbook explanation, we direct the reader to Rybicki & Lightman (1979). We capture the highlights of that discussion here.

Synchrotron radiation results from the helical motion of relativistic electrons around a magnetic field. The helical motion is a superposition of the uniform circular motion induced by the centripetal magnetic force perpendicular to the field direction and any constant velocity in the direction parallel to the field. By equating the magnetic ($q\mathbf{v} \times \mathbf{B}$) force with the centripetal force, one calculates the frequency of gyration,

$$\omega_B = \frac{qB}{\gamma m}.$$

The total power emitted is then

$$P = \frac{1}{4\pi\varepsilon_o}\frac{4}{3}\sigma_T c\beta^2\gamma^2\left(\frac{B^2}{8\pi}\right),$$

where the quantity in parentheses is the magnetic energy density.

The relativistic beaming effect as the electrons circle in the field means that each electron is only emitting in the direction of the observer over part of its orbit. Because this time is less than the period, the bandwidth of the synchrotron radiation will be broader than the monochromatic ω_B. The upshot of this is that the spectrum will extend up to some critical frequency

$$\omega_c \equiv \frac{3}{2}\gamma^3\omega_B \sin\alpha,$$

where α is the inclination angle of the electron orbit.

The spectral shape of synchrotron radiation is dependent on the spectrum of energies of the input electrons. It is usually assumed that they obey a power law distribution,

$$N(E) \propto E^{-p}.$$

This leads to a power law spectrum in the power emitted by the electrons,

$$P(\omega) \propto \omega^{-(p-1)/2}. \tag{3.71}$$

The polarization of the radiation is in aggregate linear with its direction perpendicular to the magnetic field. The maximum value for synchrotron polarization is 75%; disordering of the fields or superposition along the line of sight can decrease the observed value in an astrophysical setting.

3.9 Polarized Radiative Transfer

The main staple of understanding radiative transfer along a line of sight is the well-known radiative transfer equation that couples the change in intensity with respect to unit optical depth, $\frac{dI}{d\tau}$ to the intensity, I_ν and the source function, J_ν, which describes the addition to the beam of light at any given position along this path,

$$\frac{dI_\nu}{d\tau} = -I_\nu + J_\nu. \tag{3.72}$$

Many texts derive this equation (Rybicki & Lightman 1979), which details the change in intensity at a given point along this path, taking into account the energy removed from and that added to the beam at each point. While this is an excellent model for unpolarized light, polarization adds a complication that the effect on all four Stokes parameters must be considered along this path. This not only includes contributions to and extinction from the beam, but also interactions that can alter the state of polarization from the input of each line element to the corresponding output. However, the basic idea remains the same as for the scalar case, and the concept of Müller matrices may be employed to account for the polarization,

$$\frac{d\mathbf{S}}{d\tau} = -\mathsf{MS} + \mathbf{J} \tag{3.73}$$

$$\frac{d}{d\tau}\begin{pmatrix} I_\nu \\ Q_\nu \\ U_\nu \\ V_\nu \end{pmatrix} = -\begin{pmatrix} M_{II} & M_{IQ} & M_{IU} & M_{IV} \\ M_{QI} & M_{QQ} & M_{QU} & M_{QV} \\ M_{UI} & M_{UQ} & M_{UU} & M_{UV} \\ M_{VI} & M_{VQ} & M_{VU} & M_{VV} \end{pmatrix}\begin{pmatrix} I_\nu \\ Q_\nu \\ U_\nu \\ V_\nu \end{pmatrix} + \begin{pmatrix} J_{I_\nu} \\ J_{Q_\nu} \\ J_{U_\nu} \\ J_{V_\nu} \end{pmatrix}. \tag{3.74}$$

The diagonal elements of the Müller matrix describe the extinction of the four Stokes parameters; the off-diagonal elements describe the polarization transformation of the Stokes vector. As can be seen in Equation (3.73), radiative transfer calculations that consider the vector nature of electromagnetic waves are significantly more complicated than their scalar counterpart (Equation (3.72)), involving a system of coupled linear differential equations.

Radiative transfer codes rely on the above formalism, in which M and $\mathbf{J}$ are informed by sophisticated models of the matter's composition, geometry, and environment along the line of sight. As an example, the radiative transfer code POLARIS (Reissl et al., 2018) utilizes a Monte Carlo photon propagation simulation to model observations using 3D models (such as those produced by magnetohydrodynamics codes) as inputs. This particular code allows the user to model the physical conditions along the line-of-sight in addition to taking into account the physics of the radiative torques alignment process.

References

Adachi, S., & Kennaugh, E. 1960, ITMTT, 520
Arfken, G. 1985, Mathematical Methods in Physics (3rd ed.; Orlando: Academic)
Ashcroft, N. W., & Mermin, N. D. 1976, Solid State Physics (Philadelphia, PA: Saunders)
Barkats, D., Bischoff, C., Farese, P., et al. 2005, ApJS, 159, 1
Bischoff, C., Brizius, A., Buder, I., et al. 2013, ApJ, 768, 9
Bohren, C. F., & Huffman, D. R. 1983, Light Scattering by Small Particles (New York: Wiley)
Cataldo, G., Wollack, E. J., Barrentine, E. M., et al. 2015, RScI, 86, 013103
Chuss, D. T., Wollack, E. J., Henry, R., et al. 2012, ApOpt, 51, 197
Collin, R. E. 1990, Field Theory of Guided Waves (New York: Wiley)
Cortiglioni, S. 1994, RScI, 65, 2667
Essinger-Hileman, T. 2013, ApOpt, 52, 212
Franta, D. 2020, JAP, 127, 223101
Hagen, E., & Rubens, H. 1903, Annalen der Physik, 316, 873
Hecht, E. 2017, Optics (5th ed.; New York: Pearson)
Jackson, J. 1972, Classical Electrodynamics: Second Edition (New York: Wiley)

Joannopoulos, J. D., Johnson, S. G., Winn, J. N., & Meade, R. D. 2008, Molding the Flow of Light (Princeton, NJ: Princeton Univ. Press)
Jovanovic, N., Gatkine, P., Anugu, N., et al. 2023, JPhP, 5, 042501
Kerr, A. 1999, MMA Memo, 21, 1
Kraus, J. 1984, Electromagnetics (3rd ed.; New York: McGraw-Hill)
Leontovich, M. A. 1948, Investigations on Propagation of Radio Waves, ed. B. A. Vvedenskii (Moscow: AN SSSR) 5
Lucarini, V., Peiponen, K.-E., Saarinen, J. J., & Vartiainen, E. M. 2005, Kramers-Kronig Relations in Optical Materials Research (Berlin: Springer)
Lucyszyn, S. 2004, IPMAP, 151, 321
Morgan, S. P. 1949, JAP, 20, 352
Munk, B. A. 2005, Frequency Selective Surfaces: Theory and Design (New York: Wiley)
Ordal, M. A., Bell, R. J., Alexander, R., Long, L. L., & Querry, M. R. 1985, ApOpt, 24, 4493
Ordal, M. A., Long, L., Bell, R. J., et al. 1983, ApOpt, 22, 1099
Palik, E. 1998, Handbook of Optical Constants of Solids (Amsterdam: Elsevier)
Pancharatnam, S. 1955, PIAS, 41, 137
Pisano, G., Nesti, R., Ng, M., et al. 2012, JEWA, 26, 707
Pozar, D. M. 2012, Microwave Engineering (4th ed.; Hoboken, NJ: Wiley)
Reissl, S., Wolf, S., & Brauer, R. 2018, POLARIS: POLArized RadIation Simulator, Astrophysics Source Code Library, ascl:1807.001
Renbarger, T., Dotson, J. L., & Novak, G. 1998, ApOpt, 37, 6643
Rybicki, G. B., & Lightman, A. P. 1979, Radiative Processes in Astrophysics (New York: Wiley)
Savini, G., Pisano, G., & Ade, P. A. R. 2006, ApOpt, 45, 8907
Senior, T. B. A. 1960, ApSRB, 8, 418
Shiles, E., Sasaki, T., Inokuti, M., & Smith, D. 1980, PhRvB, 22, 1612
U-Yen, K., Rostem, K., & Wollack, E. J. 2018, ITAS, 28, 1
van de Hulst, H. C. 1957, Light Scattering by Small Particles (New York: Dover)
Von Hipple, A. 1954, Dielectrics and Waves (New York: Wiley)
Wang, B., Zhou, J., Koschny, T., Kafesaki, M., & Soukoulis, C. M. 2009, JOptA, 11, 114003
Wheeler, H. 1942, PIRE, 30, 412
Wollack, E., Jarosik, N., Netterfield, C., Page, L., & Wilkinson, D. 1993, ApL, 419, L49
Wollack, E. J. 1994, A Measurement of the Degree Scale Cosmic Background Radiation Anisotropy at 27.5, 30.5, and 33.5 GHz (Princeton, NJ: Princeton Univ. Press)
Xu, J., Lange, A., & Bock, J. 1996, Proc. of the 30th ESLAB Symp. (Paris: ESA) 69
Yariv, A., & Yeh, P. 2005, Photonics: Optical Electronics in Modern Communications (Oxford: Oxford Univ. Press)

Chapter 4

Polarimetric Instrumentation

For the design and construction of a successful polarimeter, sensitivity to the vector nature of the electromagnetic wave is a necessary but not sufficient criterion, or in the words of far-infrared polarimetry pioneer Roger Hildebrand, "Sensitivity to polarization does not a polarimeter make." Because of the vector nature of the measurement, control of systematic errors for polarimetry is more involved than for photometric imaging systems, with the former having to deal with an additional set of considerations on top of those pertinent to the latter. For example, polarimeters, like photometers, have to mitigate the deleterious effects of time-varying backgrounds that can masquerade as signals. Simultaneously, the design of a polarimeter also has to be concerned with the effect of the instrument on faithfully acquiring and reporting the polarization state of the signal.

In considering the instrumentation for measuring polarization, it is critical to think holistically, taking into account the physical components, how they work together in the system, and the strategy for data reduction. As such, this chapter begins with a description of a generic astronomical polarimeter for measurements in the far-infrared through millimeter. This polarimeter concept captures the basic required elements along with their relationship to the general strategy for mitigating systematic effects. We then describe various polarimeter components. We review some key concepts in sampling that apply to both the temporal (data acquisition) and spatial (map) domains. We follow with a detailed discussion of thermal background and noise; understanding the thermal background, whether it be from the atmosphere, telescope, or source itself, is critical for designing a system that is optimized to approach the statistical thermal noise limit for continuum systems operating in the far-infrared through millimeter. We conclude this chapter with a discussion of key concepts related to the operation and calibration of a polarimeter: instrumental polarization, isolation, efficiency, cross-polarization and polarization angle calibration.

doi:10.1088/2514-3433/ae1022ch4

4.1 A Polarimetry System

To set the stage for later discussion, we present a general picture of a polarimeter in Figure 4.1. Borrowing from the lexicon of remote sensing, we wish to measure a *scene*, which is represented by a Stokes vector that can be a function of both space and time,

$$\mathbf{s}(\mathbf{r},\, t) = [I(\mathbf{r},\, t),\ Q(\mathbf{r},\, t),\ U(\mathbf{r},\, t),\ V(\mathbf{r},\, t)].$$

The goal of the polarimeter is to produce a high-fidelity representation of the scene, which is called an *image* or *map*, which is represented as

$$\hat{\mathbf{s}}(\mathbf{r},\, t) = [\hat{I}(\mathbf{r},\, t),\ \hat{Q}(\mathbf{r},\, t),\ \hat{U}(\mathbf{r},\, t),\ \hat{V}(\mathbf{r},\, t)].$$

The scene is often viewed in the presence of foregrounds, which can contribute a combination of additional signal, or in the case of atmosphere and the emission of the optical components, some combination of statistical and correlated noise.

The hardware of a polarization system is similar to that of any continuum single-dish system in the FIR through millimeter. The *telescope* defines the aperture that ultimately sets the angular resolution of the system. The *receiver* houses the detector and defines the polarization detection parameters using a selected diplexing strategy (more about this below). It also contains elements that define the spectral range accepted by the system. The telescope, its pointing system, and receiver system are used in concert to design the modulation scheme. This is accomplished through a combination of explicit polarization modulators and/or instrument and sky rotation, depending on the global architecture implemented for the experiment. We will expand on the discussion of modulators and modulation in what follows.

The *data acquisition system* coordinates sampling of the detector and optical system carries these signals, and stores them electronically for subsequent analysis. The sensor information required to reconstruct the instrument gain, pointing, and calibration solutions are also captured. Synchronization is crucial here, as the raw data products produced by the polarimeter must allow reconstruction of the state of the telescope and polarimeter system for each data point collected. Parameters that must be tracked are system-dependent but can include (though not necessarily limited to) quantities such as boresight pointing location, array and detector rotation relative to sky coordinates, focal plane geometry, modulator state, key instrument components temperatures, sky opacity, etc. Data reduction will be covered in Chapter 6, but we will lay the groundwork in this chapter.

4.2 Telescope and Optics

In many instances, particularly in the far-infrared and submillimeter, polarimetric cameras are designed as back-end instruments to existing telescopes. Depending on observing wavelength, desired resolution, and other constraints, the diameter of such telescopes have ranged from $\sim$1 m, for the case of the Kuiper Airborne Observatory (Cameron 1976), to 50 m for the Large Millimeter Telescope (Hughes et al. 2010, 2020). Such telescopes have achieved angular resolutions down to $\sim 5''$.

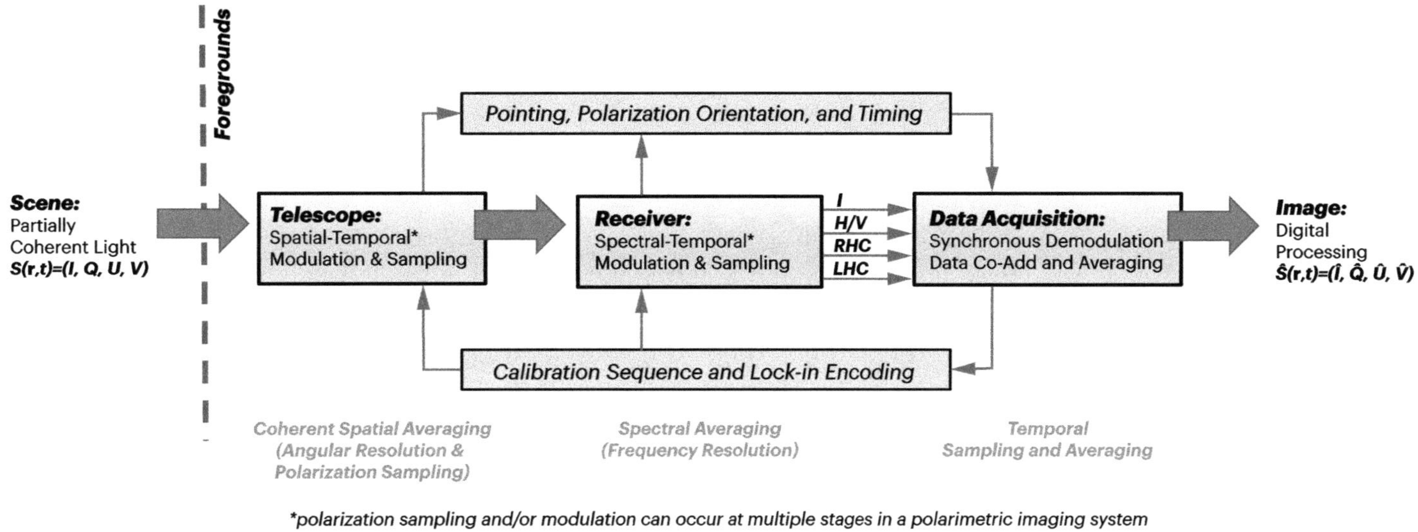

Figure 4.1. The components of a generic polarimeter are shown. Light comes into the polarimeter either from the telescope or the sky (in the case where the polarimeter is integrated into the telescope). The light experiences polarization modulation, either as a results of the observing scheme or via an explicit modulator. Finally, the light is separated into two orthogonal polarization states via a diplexer, and one or both states are measured by a detector. This can be done either using linear polarization states ($\pm Q$, $\pm U$, or a linear combination thereof) or circular polarization states ($\pm V$).

This corresponds to a spatial resolution of ~0.01 pc for local clouds such as the Orion Molecular Complex and ~0.2 pc at the Galactic center.

For longer wavelengths, observatories are sometimes constructed with the main purpose of measuring the cosmic microwave background and the intervening foregrounds. In these cases, the telescope design can be tailored to match the relevant science goals, often requiring angular resolutions of several arcminutes. For example, the BICEP (Keating et al. 2003) series of telescopes utilizes on-axis lens-based optical assemblies. The entire telescope is cooled to mitigate background noise and increase thermal stability. The on-axis design allows ease of rotating the telescope to change the linear polarization sensitivity. Each refractive optics tube in the telescope array is unobstructed and achieves low instrumental polarization. Most reflective telescopes built for CMB polarimetry utilize off-axis designs because diffraction is significant at these long wavelengths relative to their optical counterparts, and thus it is advantageous to avoid placement of the secondary mirror within the optical path upstream of the primary. The Gregorian design can accommodate large focal planes and is utilized in such telescopes as ACTpol (Koopman et al. 2016) and the South Pole Telescope (Ruhl et al. 2004). In such design implementations it is possible to use the interplay between primary and secondary mirror to reduce cross-polarization effects geometrically (Dragone 1982; Niemack 2016).

Finally, cases in which instruments use front-end modulation schemes have required telescopes for which the entrance pupil is located not at the primary powered optic but at the modulator, which is positioned at the front of the instrument. The Primordial Inflation Polarization Explorer (PIPER) (Chuss et al. 2010) and Cosmology Large Angular Scale Surveyor (CLASS) (Essinger-Hileman et al. 2016) are two examples of this design.

Many back-end cameras require re-imaging optics. Lenses can produce polarization in an initially unpolarized light (essentially $T \to P/I \to P$ mixing) (Pisano 2003). Similarly, grids (Lazareff et al. 2009), reflective surface, and scattering can lead to spurious polarization artifacts. Ideally, impact of such instrumental polarization sources should be assessed, allocated, and mitigated as needed to achieve the desired level of systematic control consistent with the needs of the end application.

The importance of stray light control in polarimetric instrumentation and measurement cannot be overstated. The performance of high sensitivity systems at wavelengths longward of the infrared can be severely degraded by the sidelobe response, diffraction, scattering, and self-emission. An excellent review of stray light mitigation strategies is presented in Pompea (1995). Generally, these effects can lead to spurious polarized signatures, glint, and excess noise. Design choices made during the earliest stages of the instrument concept development ultimately influence the system's capability and potential control over systematic artifacts in data products. This said, stray light analysis, testing, and model validation of the system implementation also play an important role though commissioning and end astrophysical use. These considerations highlight the importance and challenges of "stray light management" from the perspective of delivering the envisioned technical capability in a cost-effective manner. In identifying critical elements of the optical the design, it is crucial to carefully consider what the detector sees. This involves looking beyond the beam-defining optical elements and stops for unintended secondary effects and light paths. In practice, high performance systems have many stray light issues that are not important in lower performance systems. Be diligent.

The electromagnetic properties of instrument elements are important to characterize and understand in detail. The absorptive properties of materials and their roughness relative to a wavelength play important roles in the implementation of optical baffles, stops, and light trap structures for termination of radiation in calibrators and optical systems. A useful starting point for "black" absorber coatings and surface treatments is the review by Persky (1999). Several absorptive (Leinert & Klüppelberg 1974; Bock et al. 1995; Gervasi et al. 1998) and reflective (Rock et al. 1982) baffle design implementations are insightful examples to study in detail. For additional baffle and scattering considerations encountered with metamaterial and mesh filters, see Sharp et al. (2012). The baffle analysis approach presented in Haghshenas & Johari (2014) and with refinements by Rohrbach (2023) allows an exploration of the design and performance trades between baffle angular acceptance, blade geometries, and total number of baffle blades.

Other stray light control techniques commonly encountered in electrically large aperture antenna and telescope settings include: (1) co-moving reflective screens to block radiation and reduce coupling to the telescope aperture, (2) fixed reflective "ground screens" for terrestrial experiments, "Sun-shields" for independently pointed space borne telescope systems, etc., (3) co-moving enclosures to reduce coupling through blockage (Crawford et al. 1961), diffusive scattering (Gullett et al. 2023), absorption (Coleman et al. 2003; Datta et al. 2024), or a combination thereof, and (4) if necessary and present, adapting the shape and treatment of aperture blockages such as supporting legs (Lawrence et al. 1994; Landecker et al. 1997) and/or illuminated edges to maximize the diffractive response in the forward direction (Mather 2003; Burnside & Chuang 2003). From a sensor perspective, the primary objective of these approaches is to limit the effective field of view, to terminate residual optical response on cold sources (i.e., on sky or cooled elements), and to limit optical coupling to radiation from warm structures that are either part of the instrument or the observatory environment.

4.3 Polarization Separation

A key function within a polarimeter is that of separating the incident radiation into its polarization components. This is usually done in such a way that the two separated components are orthogonal polarization states, or states that are represented by points located 180° from one another on the Poincaré sphere. Such decomposition has the advantage that any general polarization state can be represented as a sum of these two states, so long as the phase relationship between the states is know. Though any pair of orthogonal states provides a possible solution for incoherent systems, in practice, diplexers tend to separate two linear orthogonal states (i.e., two states with electric field vectors having a relative orientation of 90°), or left-handed and right-handed circular polarization states.

Polarization diplexers can be used in a variety of ways in an instrument. In some cases, the system preserves the relative phase between the two orthogonal polarization states. If this is the case, then the full Stokes polarization state of the incoming light is recoverable. An example of this type of system is the polarized beam splitter of a Martin–Puplett interferometer. The full set of Stokes parameters of the incoming beam is nominally preserved as long as the relative phase between the two beams is tracked. A diplexer can also be used to add polarization sensitivity to a direct detector. In this case, relative phase information is forfeited, and the only Stokes parameter that is recoverable is the one that matches the basis of the polarization separation.

Figure 4.2 (adapted from Kusaka et al. 2014b) shows a variety of techniques for diplexing the polarization signal with modern detection techniques. In the left column, *single mode* detection schemes are shown. S1 shows incoherent detectors that use a feedhorn/antenna to filter the input to a single mode. Polarization sensitivity is achieved via a quasioptical diplexing grid that directs the reflected polarization to one copy of the horn-detector system and the other (orthogonal/transmitted) polarization to another copy. S2 is similar in its use of a feedhorn/antenna to define the modal content. In this case, polarization sensitivity is handled within the absorber itself, and the two orthogonally-oriented polarization-sensitive detectors make this a much more compact design. Finally, S3 shows an architecture in which the signal from the feedhorn is coupled to a on-chip photolithographic circuit that handles the beam diplexing. In this topology, the planar chip could also be a waveguide OMT (to define a linear basis) or a septum polarizer (to define a

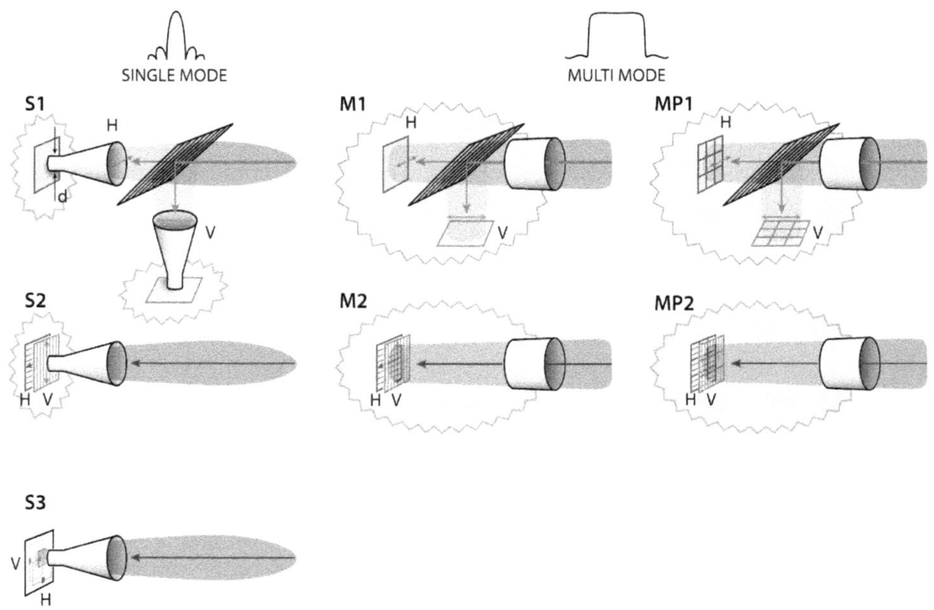

Figure 4.2. Various architectures for polarization diplexing are shown. This figure was adapted from Kusaka et al. (2014b). See text for a detailed explanation.

circular basis) for the detectors. In these figures the "feedhorn" symbol can also be interpreted as a planar antenna coupling structure used in conjunction with a dielectric immersion lens. Many CMB instruments rely on focal planes filled with variations on these types of single mode sensor architectures (Essinger-Hileman et al. 2016; Koopman et al. 2016; Ruhl et al. 2004).

The right side of the figure depicts four different types of multimode detectors, with the fore-optics represented by the tapered cylinder. M1 and M2 represent architectures with single multi-moded detection elements. In this case, the device is sensitive to all of the étendue defined by the foreoptics. M1 utilizes a wire grid polarizing diplexer; whereas, M2 shows the case where the polarization sensitivity is defined in the absorbing structure of the detector element itself. MP1 and MP2 are multi-pixel versions of M1 and M2. MP1 is representative of the HAWC+ architecture (Harper et al. 2018).

4.3.1 Free-standing Wire Grids

One of the simplest polarization diplexer technologies used from the far-infrared through microwave is the wire grid polarizer, which consists of an array of parallel metal wires. For light having wavelength much larger than the wire diameter, the surface is a metamaterial that reflects the linear polarization parallel to the wire direction and transmits the orthogonal polarization state.

The electrical properties of a polarizing grid are dependent on the material properties and geometry of the wires. Roughly speaking, one can model the interaction of each of the two polarizations with an equivalent circuit as shown in Figure 4.3. Such models and their parameter values have been solved (Marcuvitz 1951). The mode with polarization parallel to the wires can be modeled as an inductive circuit as shown in the top of the figure. The induced current from the electric field is predominantly parallel to the wire in this case. Loss is modeled via a resistor in series with this inductance. Symmetric series capacitors represent the capacitance across the width of the wires. Defining the grid *pitch* as g and the wire radius as a, (Marcuvitz 1951) gives the circuit parameters as

$$X_{\mathrm{LC}} = \frac{g}{\lambda}\left(\frac{2\pi a}{g}\right)^2 \tag{4.1}$$

$$X_{\mathrm{L}} = \frac{g}{\lambda}\left\{\ln\frac{g}{2\pi a} + \sum_{m=1}^{\infty}\left(\frac{1}{\sqrt{m^2-\left(\frac{g}{\lambda}\right)^2}} - \frac{1}{m}\right)\right\}. \tag{4.2}$$

For the orthogonal mode, a capacitive model can be used. In this case, the capacitance between adjacent wires is the dominant term, and resistive loss is negligible. The circuit is shown in the bottom half of Figure 4.4. Again, the circuit element values can be

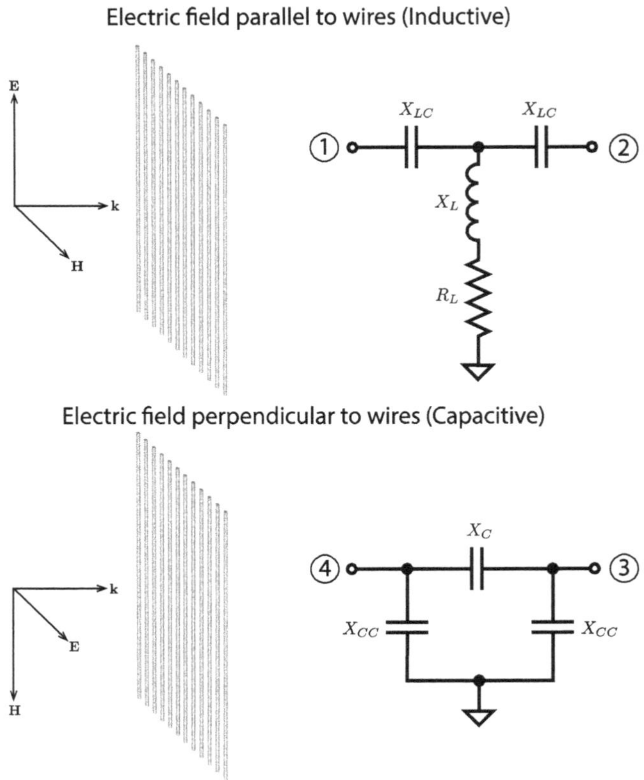

Figure 4.3. Plane waves incident on a polarizing grid can be modeled via a pair of circuits, an inductive circuit for the polarization having the electric field parallel to the wires and a capacitive circuit for the case where the electric field is perpendicular to the wire direction. The circuit parameters depend on the material properties and geometry of the grid (see main text for details).

Figure 4.4. A wire grid is shown that is used for the CLASS CMB polarization modulators (VPMs).

found from Marcuvitz (1951). Such equations have been used by Chuss et al. (2012a) to model the grid component of variable-delay polarization modulators,

$$X_{CC} = -\frac{\lambda}{g}\left\{2\left(\frac{g}{2\pi a}\right)A_2\right\} \tag{4.3}$$

$$X_C = -\frac{\lambda}{g}\left\{2\left(\frac{\lambda}{g}\right)^2\left(\frac{g}{2\pi a}\right)^2 A_1 - \frac{1}{4}\left(\frac{2\pi a}{g}\right)^2\frac{1}{A_2}\right\}^{-1} \tag{4.4}$$

Here,

$$A_1 = 1 + \frac{1}{2}\left(\frac{2\pi a}{\lambda}\right)^2\left\{\ln\frac{g}{2\pi a} + \frac{3}{4} + \sum_{m=1}^{\infty}\left(\frac{1}{\sqrt{m^2 - \left(\frac{g}{\lambda}\right)^2}} - \frac{1}{m}\right)\right\} \tag{4.5}$$

$$A_2 = 1 - \frac{1}{2}\left(\frac{2\pi a}{\lambda}\right)^2\left\{\ln\frac{g}{2\pi a} - \frac{11}{4}\right\} \tag{4.6}$$

$$+\left(\frac{2\pi a}{\lambda}\right)^2\left\{\frac{1}{24} - \sum_{m=1}^{\infty}\left(m - \frac{1}{2m}\left(\frac{g}{\lambda}\right)^2 - \sqrt{m^2 - \left(\frac{g}{\lambda}\right)^2}\right)\right\}. \tag{4.7}$$

As mentioned before, generally, wires much smaller than the wavelength are desired; however, practical considerations can constrain the design. For a given wire diameter, a useful "rule of thumb" is to set the wire pitch such that $2a/g \sim 1/\pi$. This provides a reasonable trade for the wires being sufficiently close to reflect the inductive mode, while being sufficiently far apart to transmit the capacitive mode (Chuss et al. 2012a).

Free-standing wire grids are typically made of tungsten wire glued or soldered (if they are coated with gold or copper) to a frame. Tungsten is typically chosen due to a combination of its favorable strength-to-mass ratio and its reasonably high conductivity. Using wire-wrapping techniques, it is possible to manufacture free-standing wire grids with wire diameters down to $\sim$10 μm, which provides good diplexing performance up through $\sim$few THz (Barney & Magner 1988; Eichhorn & Magner 1986). At frequencies higher than this, photolithographic grids are a preferred diplexing solution.

In the millimeter part of the spectrum, where wire diameters for diplexers can be as large as $\gtrsim$50 μm, gold or copper coatings (typically applied over a nickel or titanium interface layer) can be implemented to increase the conductivity; however, ensuring the integrity of such a coating on such small wire can be challenging. Other materials such as stainless steel (Novak et al. 1989b) and molybdenum (Lahtinen & Hallikainen 1999) have been implemented for particular applications.

The manufacture of wire grids is typically done by wrapping wire from a spool of drawn wire onto a frame. Early efforts typically utilized a lathe or milling machine to tumble a pair of grid frames, while a wire tensioner feeds the wire onto the frames. The wires are then attached to the frames via epoxy and the frames are separated producing a grid pair (Martin & Puplett 1970; Costley et al. 1977; Chambers et al. 1986). Improvements have been made to this process to improve wire spacing uniformity.

A modification of wire wrapping has been developed (Novak et al. 1989b) in which the wire is initially wrapped on a cylindrical mandrel that has a pair of removable inserts that runs along the length of the cylinder. The cylinder is wrapped with wire, and the wire is glued to the inserts. The wire between the inserts is cut and the grid is removed from the mandrel. This technique was initially for a 25 mm $\times$ 25 mm grid prototype using 10 μm diameter stainless steel wires for THz applications. It was modified for CMB applications (Voellmer et al. 2008) to create a $\lesssim$1 m grid using gold-plated tungsten wires. In this case, grooves were cut into the mandrels before wrapping to ensure uniform wire placement. Subsequent implementations of this technique (Chuss et al. 2014; Harrington et al. 2021) used the edge of an end mill to cut grooves in the insert only.

4.3.2 Photolithographic Polarization Diplexers

A similar technology to free-standing wire grids consisting parallel planar wires is to create a similar pattern on a thin low-loss dielectric substrate (Bird & Parrish 1960; Auton 1967). The geometry of the conductors in this case is slightly different; however, on an electrically-thin low-index supporting membrane the physics is similar. Such grids are used in the HAWC+ instrument (Harper et al. 2018) as the polarization beam splitter. In this case, they are produced as 1 μm wide stripes with a 2 μm pitch on a 0.9-to-1.5 micron thick mylar film (QMC Instruments model “P2”). Polarizers of this general construction are also found in mid-to-far-IR polarizing spectral-interferometers (Crooke 1998) and employed as polarization calibration sources.

4.3.3 Orthomode Transducers (OMTs)

Orthomode transducers (OMTs) are linear polarization diplexers realized in waveguide. These passive devices operate in a regime where a limited number of discrete modes are accessible and defined by the waveguide geometry. In this setting symmetry plays an important role in defining the electrical performance and the achievable spectral range of the device. In comparing to a quasi-optical or free-space wiregrid polarizer, which are inherently broadband and operate on the detector array scale, an OMT can be considerably more compact, operating at the polarized detector level. These differences have led to the adoption of grid polarizers in the far-IR and OMTs in the microwave, respectively, for diplexing polarization sensitive focal planes.

These waveguide structures are inherently 4-port devices having a common port to simultaneously receive two orthogonal polarizations and a pair of single mode ports that provide the separated signals for further processing. See Figure 4.5 for schematic representations of such 4-port devices for linear and circular polarizations. Examples include: (1) narrow-band asymmetric orthomode transducers for linear polarization separation, which are typically limited by the excitation of higher

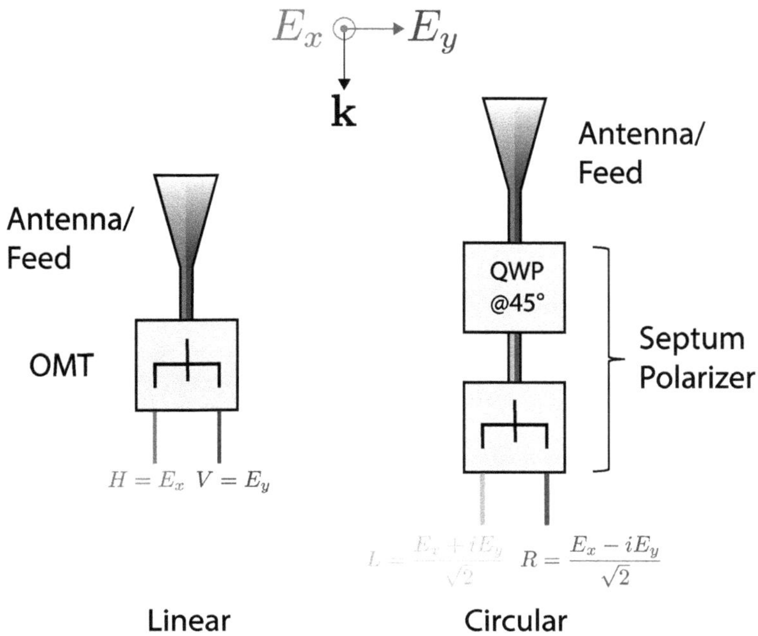

Figure 4.5. Examples of 4-port OMT structures for linear (LEFT) and circular (RIGHT) polarization separation. In each case, the "Antenna/Feed" refers to the common port that contains both polarizations, and the separated polarizations exist in two single mode ports at the bottom of each figure.

order modes in the common-mode part, (2) narrow-band septum polarizers for circular polarization, (3) full-band 4-fold symmetric orthomode transducers for linear polarization achieve high polarization isolation. Symmetric OMT structures are topologically related to (or derived from) turnstile junctions and quadridge orthomode transducers. These devices are commonly implemented as metallic split-block or electro-formed waveguide structures (Uher et al. 1993).

Planar four-fold symmetric orthomode transducers have been enabled through the use of low-loss dielectrics and superconducting transmission lines as described in Wollack (2009) and McMahon et al. (2009). See Figure 4.6 for a simplified representation. Since its introduction, this planar polarization coupling architecture has been widely-adopted and used to realize wide-band detector arrays with high coupling efficiency. These structures are fabricated using photolithographic and silicon micro-machining tools and have on-chip polarization separation, spectral band-pass definition, and integrated sensors. In this detector coupling approach the sensors typically resides within a metallized enclosure defined by micro-machined silicon (Crowe et al. 2012), which can be channelized to limit high frequency radiation from reaching the sensor. Symmetric HE_{11} feed structures realized via machining or platelets stacks (Gundersen & Wollack 2009) and related smooth-wall structures (Zeng et al. 2010) have found use in defining the sensor's illumination function from bulk metal and/or silicon. Photonic choke joints can be used to realize a low-loss and relatively dimensionality-insensitive interface between dual-mode waveguide and the feedhorn structure (Wollack et al. 2010). Sinuous (O'Brient et al. 2010), dual-slot (Chattopadhyay & Zmuidzinas 1998), and arrayed antenna

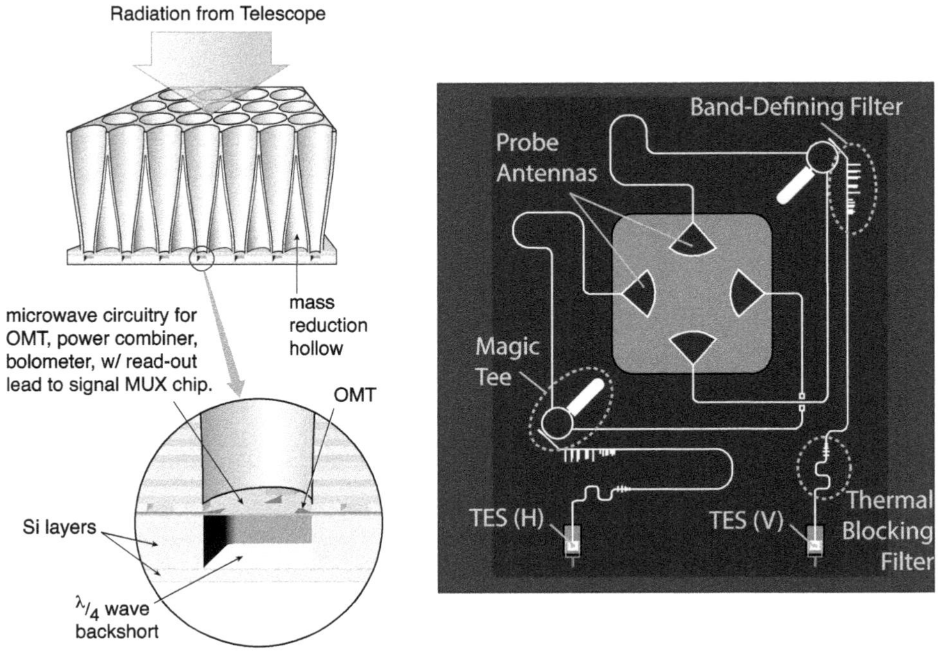

Figure 4.6. A feedhorn-coupled planar OMT schematic is shown from (Chuss et al. 2016). (Left) A dual-mode antenna (feedhorn) guides radiation onto a photolithographed silicon device containing the planar OMT and detectors. (Right) The planar OMT chip layer is shown. The pairs of antenna probes separate the two linear orthogonal polarizations from the feedhorn, appropriately combine each signal in a magic-tee, and guide each signal to a network of filters and a detector (TES).

structures (Kuo et al. 2008) provide other examples of planar devices, which functionally serve this instrumentation role where the antenna structures are directly integrated on the silicon wafer (i.e., rather than by vertically stacking layers).

4.4 Polarization Modulation

Polarization modulation is the encoding of the polarized signal of interest in the time-ordered data in a way that allows one to completely characterize the desired Stokes parameters of the signal and to distinguish the polarized component from some combination of the associated unpolarized component of the source and any systematic signals that can be introduced through the environment and detection scheme. Subsequent demodulation is then employed to separate the effects of the polarized signal from the larger, unpolarized signal and any relevant systematics.

There are two main reasons why polarization modulation is important. The first is that characterization of the complete polarization state of the incoming light requires that the state to which the instrument is sensitive be changed, or modulated, in some way. Second, particularly for astronomical signals in the range of the electromagnetic spectrum considered in this book, the polarized component of the signal is much smaller than both the total intensity of the source and the in-band backgrounds radiation. In essence, polarization modulation enables variations in signal from a polarization-sensitive detector that are due to true polarization from the source that is being measured from other fluctuations. Such spurious fluctuations can include gain and responsivity variations, variations in loading on the detector, etc.

Modulation as defined in this general sense is necessary for any polarization measurement. It can either be implemented as an explicit instrument element (often referred to as a "modulator"), such as a half-wave plate, or it can be done as part of the observation strategy. For example, a combination of sky rotation, instrument rotation, and careful calibration of polarization-sensitive detectors can be used to separate the desired polarization from other components of the time stream.

For polarimeter design, modulation strategy is a key consideration. Modulator decisions are highly-coupled to the observation strategy that will be used. To start the discussion, we will discuss general properties of a modulator. Later, we will discuss specific techniques used to modulate polarization along with their practical implementations.

Polarization modulation refers to the procedure by which the polarization state of light entering the modulator is systematically transformed into other polarization states. Equivalently, one can interpret polarization modulation as systematically changing the polarization state to which the instrument is sensitive. The advantage of doing this is that because the transfer function of the modulator is known, it can be synchronously demodulated, and the polarization signal can be separated from the unpolarized flux that is generally much larger than the polarized signal.

Generally, any device that changes the state of the input polarization can be used to modulate a polarization; however, it is generally appreciated that modulators that decouple the polarized and unpolarized flux are preferred. Mathematically, this statement is equivalent to the preservation of the polarization which, as introduced in Chapter 2, is defined by

$$P^2 \equiv Q^2 + U^2 + V^2, \tag{4.8}$$

where again, Q, U, and V are the Stokes parameters representing linear and circular polarization, respectively.

Systems that preserve P are described by Jones matrices that are unitary. These matrices constitute the group of similarity transformations on the Poincaré sphere. Equivalently, such transformations are described by the group of Müller matrices that are real and orthogonal (Sternberg 1994). The Müller matrices that describe ideal polarization modulation are represented by rotations in the space of the Poincaré sphere. This is mathematically related to the fact that there is a homeomorphism between 2×2 special unitary matrices, SU(2) and the group of three-dimensional rotations, O(3).

There are ways to achieve polarization modulation outside of this restriction. For example, one could rotate a polarizer in front of a polarization-sensitive detector. Because such systems also modulate Stokes I, the power loading on the detectors is also significantly modulated, leading to variations in detector response that can significantly complicate the polarization determination and the instrument calibration in general.

4.4.1 Ideal (Lossless) Modulator Topologies

Unitarity for a modulation Jones matrix represents an ideal condition. Real modulators of course have more complexity. For example, loss can couple I and P. Wavelength dependence in phase delay can limit the useful bandwidth of a device, and this effect is dependent on specific implementation. In this section, we will discuss ideal modulation topologies, which we will define as those that can be described by unitary Jones matrices. We will then discuss implementations based on these topologies and give some examples of deviation from ideal performance along with some of the mitigation strategies that have been applied.

Physically, the decoupling between I and P is characteristic of systems that are lossless. Such instruments are, of course, impossible to realize in practice, but it is useful to differentiate systems that mix I and P as a result of their general architecture from those that do so because of second-order imperfections. In this section, we discuss a set of commonly used modulation strategies that provide starting points for real-world modulators.

The unitary condition for modulation is uniquely realized by the introduction of a phase delay between two orthogonal polarizations. An example of Jones and Müller matrix development for a phase delay between two linear orthogonal states was shown in Chapter 2. It was shown that this led to a rotation about the axis connecting the two applicable polarization states and that the magnitude of the rotation was equal to the introduced phase delay. It follows that polarization modulation is well-parameterized in terms of two variables: one representing the basis state of the polarization and one representing the magnitude of the phase delay. Either (or both) of these parameters can be varied to produce polarization modulation.

We can describe an ideal polarization modulator by a general unitary matrix,

$$\mathbf{J}_{\text{mod}} = \begin{pmatrix} \alpha & \beta \\ -\beta^* & \alpha^* \end{pmatrix}. \tag{4.9}$$

We note that this matrix is a representation of a general member of SU(2), the group of 2×2 complex matrices with determinants of $+1$. For simplicity, we use this form for the following analysis, noting that an analogous calculation can be done for those unitary matrices with determinants of -1 with the same results.

The α and β elements in the Jones matrix representation of the polarization modulator can be parameterized by variables that can be systematically changed in time. We suppress this parameterization in the following expressions for simplicity. As such, a general polarization modulator coupled to a pair of detectors that are sensitive to two orthogonal polarization states will detect power in each detector that is a function of the input Stokes parameters and the modulation matrix elements,

$$\mathscr{P}_{\pm} = \frac{1}{2}\{I \pm [Q(\alpha^2 - \beta^2) + 2U\Re(\alpha\beta^*) + 2V\Im(\alpha\beta^*)]\}.$$

4.4.2 Introduction of Loss

Next, we examine systems in which the polarization modulation is unitary, but the system in general is not. Specifically, we consider the dependence of the position of the modulator in the signal chain in its capability to mitigate systematic errors.

Figure 4.7 shows two such systems that differ in the location of the polarization modulator relative to the loss (gain) in the system. In each system, losses are modeled by gain terms, g_+ and g_-. In practice, these differential gain terms could represent asymmetry in the transmission lines that carry the two polarizations, differential losses in the two polarizations reflecting off of an off-axis mirror, or the effect of differential beam patterns in each of the two polarizations.

In case A, the polarization modulator is placed between the gain terms and the detectors, and for simplicity, the gain terms correspond to the H and V polarization components. The transformation matrix for each detector is given by

$$\begin{aligned}\mathscr{P}_\pm = \frac{1}{2}(g_+^2 + g_-^2)I + \frac{1}{2}(g_+^2 - g_-^2)Q \pm \frac{1}{2}\Big\{I\big[g_+^2(2\alpha^2 - 1) + g_-^2(2\beta^2 - 1)\big]+ \\ Q[g_+^2(2\alpha^2 - 1) + g_-^2(2\beta^2 - 1)] + 2U\Re(g_+g_-^*\alpha\beta^*) + 2V(g_+g_-^*\alpha\beta^*)\Big\}.\end{aligned} \tag{4.10}$$

In this case, the unitarity of the system is broken, which causes mixing of Stokes I into the polarization. The different gains are modulated and spurious polarization signals appear in the demodulated detector signal.

Case B shows a modulator placed in the front of the polarimeter's optical path. For this configuration, the signal in each of the detectors is

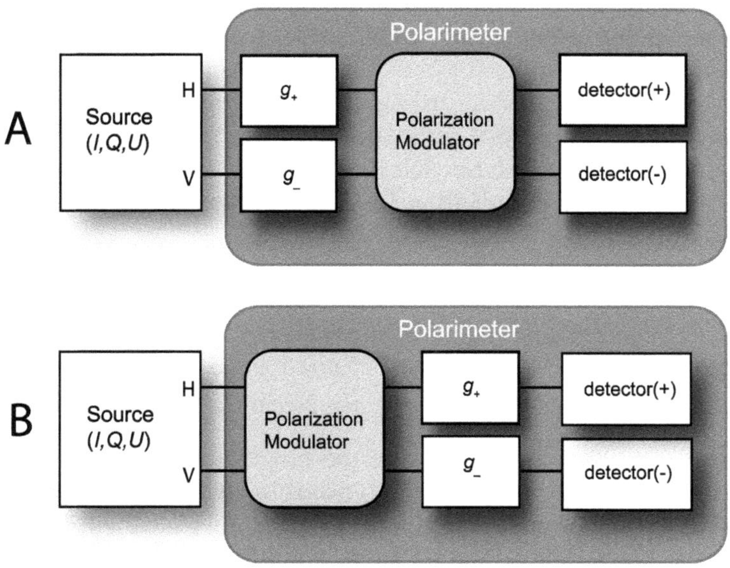

Figure 4.7. Schematic representations of two polarization architectures are shown; the difference between the two is in the relative position of the polarization modulator. In (A), the polarization modulator is after gain asymmetry (represented by g_+ and g_-). In (B), the modulator comes before differential loss in the signal chain.

$$\mathscr{P}_{\pm} = \frac{g_{\pm}}{2}\{I \pm [Q(\alpha^2 - \beta^2) + 2U\Re(\alpha\beta^*) + 2V\Im(\alpha\beta^*)]\}. \tag{4.11}$$

In this case, the asymmetry in the system merely enters as a separate gain for each detector and does not induce a mixing between I and P. From an architectural standpoint, this motivates the placement of the polarization modulator as far forward in the optical system as possible. In this case, then, one mitigates the non-unitarity of the rest of the optical system and must only be concerned with the deviation from unitarity (systematics) of the modulator itself. In real polarimeter design, other practical considerations and trades come into play that may outweigh this consideration.

4.4.3 Polarization Modulators—General Discussion

As mentioned above, a device that transforms the polarization state in a way that (at least ideally) can be represented by a unitary Jones matrix (or an orthogonal Müller matrix) are called *modulators*. As we saw earlier in this book, this is uniquely represented by an introduction of a phase delay between two orthogonal polarization states. Such a transformation corresponds to a rotation on the Poincaré Sphere about an axis that which connects the two orthogonal polarization states between which the phase is introduced. The magnitude of the rotation is equal to the magnitude of said phase.

These two states can be any two orthogonal states, which are represented by opposite directions on the Poincaré sphere; however, it is most common to introduce this phase either between two orthogonal *linear* polarization states, which are represented by any pair of opposite points on the equator of the Poincaré sphere, or the two circular polarization states, which are located at the two poles. There are several reviews in the literature (Pisano et al. 2014; Abitbol et al. 2017).

We can parameterize a general orthogonal transformation by a physical angle θ that represents the angle of basis separation (relative to some fiducial orientation) and the magnitude of the phase delay, δ. For a modulator utilizing a phase introduction in the linear polarization basis, the general Müller matrix is (Novak et al. 1989a),

$$\mathbf{M} = \begin{pmatrix} 1 & 0 & 0 & 0 \\ 0 & \cos^2 2\theta + \sin^2 2\theta \cos\delta & \cos 2\theta \sin 2\theta(1 - \cos\delta) & \sin 2\theta \sin\delta \\ 0 & \cos 2\theta \sin 2\theta(1 - \cos\delta) & \cos^2 2\theta \cos\delta + \sin^2 2\theta & -\cos 2\theta \sin\delta \\ 0 & -\sin 2\theta \sin\delta & \cos 2\theta \sin\delta & \cos\delta \end{pmatrix}. \tag{4.12}$$

For modulators to systematically change the polarization in a controlled way, at least one of these degrees of freedom (θ, δ) needs to be controllably varied. In most simple modulators, one of these is held fixed and the other is varied. In the following, we first describe types of modulators from a mathematical perspective. We then proceed to provide examples of the implementation of these various schemes. Finally, we discuss deviations from the ideal unitary/orthogonal performance.

4.4.3.1 Fixed Phase Delay Modulators

One modulation strategy is to introduce a constant phase between two orthogonal polarization states and then vary the polarization states. Because there is only one set of orthogonal circular polarization states, the solution for a constant phase delay between circular polarization states is incapable of modulation by changing basis. Thus, it is only useful to discuss modulators in this category for which a fixed phase is introduced between orthogonal linear states. The basis is then systematically changed. Implementations of fixed phase delay modulators are typically designed to approximate phase delays of either π or $\pi/2$.

An example of the former is the *half-wave plate*. The Poincaré topology for the half-wave plate is shown in Figure 4.8A. For a linear polarization, which always lies on the "equator" of the Poincaré sphere, a phase delay of π corresponds to what is effectively a reflection around the axis of phase separation (also located in the

A

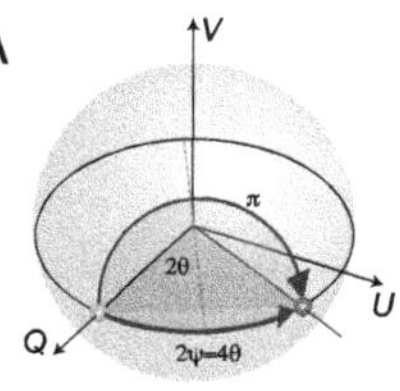

Topology: π phase delay between linear states; basis of phase separation is rotated
Free-space Example: Rotating Half-wave plate
Waveguide Example: Rotating Waveguide & π differential phase shifter

$$\begin{pmatrix} 1 & 0 & 0 & 0 \\ 0 & \cos 4\theta & -\sin 4\theta & 0 \\ 0 & \sin 4\theta & \cos 4\theta & 0 \\ 0 & 0 & 0 & 1 \end{pmatrix}$$

B

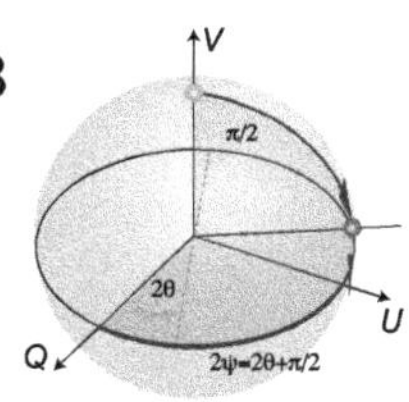

Topology: $\pi/2$ phase delay between linear states; basis of phase separation is rotated
Free-space Example: Rotating Quarter-wave plate
Waveguide Example: Rotating Waveguide & $\pi/2$ differential phase shifter

$$\begin{pmatrix} 1 & 0 & 0 & 0 \\ 0 & \cos^2 2\theta & -\frac{1}{2}\sin 4\theta & -\sin 2\theta \\ 0 & \frac{1}{2}\sin 4\theta & -\sin^2 2\theta & \cos 2\theta \\ 0 & \sin 2\theta & -\cos 2\theta & 0 \end{pmatrix}$$

C

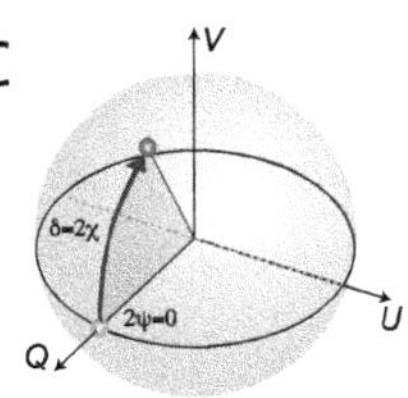

Topology: variable phase delay (δ) between linear states; basis of phase separation is fixed.
Free-space Example: Variable-delay Polarization Modulator
Waveguide Example: $\pi/2$ differential phase shifter & Faraday Rotator with delay δ

$$\begin{pmatrix} 1 & 0 & 0 & 0 \\ 0 & \cos\delta & 0 & -\sin\delta \\ 0 & 0 & 1 & 0 \\ 0 & \sin\delta & 0 & \cos\delta \end{pmatrix}$$

D

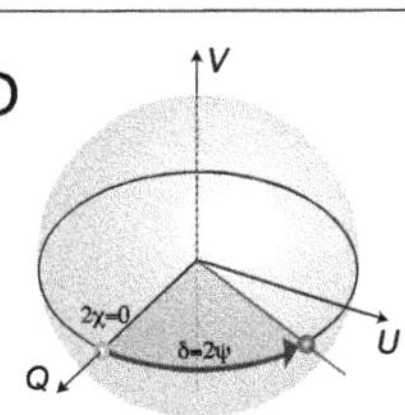

Topology: variable phase delay (δ) between circular states; basis of phase separation is fixed.
Free-space Example: Translational Polarization Rotator (TPR)
Waveguide Example: Faraday Rotator with delay δ

$$\begin{pmatrix} 1 & 0 & 0 & 0 \\ 0 & \cos\delta & -\sin\delta & 0 \\ 0 & \sin\delta & \cos\delta & 0 \\ 0 & 0 & 0 & 1 \end{pmatrix}$$

Figure 4.8. Four general modulation schemes are shown here. In each case, the basis of polarization separation is shown as a dotted green line. The magnitude of the phase delay in (A) and (B) is fixed at π and $\pi/2$, respectively. In these cases, modulation is achieved by changing the basis. In (C) and (D), the phase delay is given by δ, and the basis is fixed. Figure based on (Chuss et al. 2012b).

horizontal plane). Such a configuration is shown in the figure. The Müller matrix is also shown, which is effectively a rotation with respect to the angle 4θ where 2θ is the angle between the Q-axis and the axis of separation in Poincaré space. The angle θ is the angle in real space between the Q-direction and the angle of the basis separation. For time-variable modulation, the angle θ is systematically changed via the physical rotation of the wave plate. Because half-wave plates are widely used as polarization modulators for linear polarization from the optical through the millimeter, we will discuss implementations in more depth later in this chapter.

The *quarter-wave plate* is an example of a $\pi/2$ fixed-delay device. In this case, illustrated in Figure 4.8B. A phase delay of $\pi/2$ is introduced between two orthogonal linear polarizations, which lie at the opposite ends of the green dotted line located in the Q-U plane and oriented at an angle of 2θ with respect to the Q-axis. Because quarter-wave plates are often used when measurement of circular polarization is desired, we illustrate the operation of the modulation topology on a circular polarization state (Stokes $+V$). The $\pi/2$ phase delay rotates the $+V$ state onto the equator of the Poincaré sphere at a point 90° from the axis of polarization separation. That is, the quarter-wave plate converts circular polarization to linear polarization. This is useful because often it is easier to construct detectors that are sensitive to linear polarization than circular. At each angle of rotation of the plate, the circular polarization is converted to a *different* linear polarization, thus creating a variation of the signal on the detector.

4.4.3.2 Modulators That Use a Variable Phase Delay

Alternately, polarization modulation can utilize a constant basis of polarization separation and vary the phase delay. Here, we can consider phase introductions between both linear and circular polarization states, as the character of the modulation is different in each of these cases.

One way to realize this type of modulators is to implement the phase delay between linear polarization states. In this case, the path of modulation is a great circle that includes the poles of the Poincaré sphere (the $\pm V$ state). This is shown in Figure 4.8C for a phase separation basis that is aligned with Stokes U. The Müller matrix for such a transformation is simply a rotation between Stokes Q and V, with no change in U. Variable-delay polarization modulators (VPM) (Chuss et al. 2006, 2012a) utilize this topology.

A second type of variable phase delay modulation can occur when the phase is introduced between the two circular polarization states. In this case, shown in Figure 4.8D, the path of modulation is along the equator, and the corresponding Müller matrix is a rotation between the Stokes Q and U parameters with V fixed. Examples of modulators utilizing this type of modulation include Faraday rotators (Moyerman et al. 2013) and translational polarization rotators (TPRs; Chuss et al. 2012b).

4.4.4 Wave Plates

The most common examples of fixed-phase modulation are wave plates. We will focus on half-wave plates (π phase delay) here, since most astrophysical phenomena for the wavelengths under consideration involve linear polarization, however, quarter-wave plates ($\pi/2$ phase delay) which are encountered in dealing with circular polarization are also widely used. Wave plates are quasioptical devices. Though analogs exist that provide the same functionality in waveguides or other guided-wave systems, we will mostly concern ourselves here with free-space systems.

4.4.4.1 Birefringent Plates

The basic canonical wave plate implementation consists of a slab of birefringent dielectric material. *Birefringence* refers to the material property whereas the index of refraction is polarization-dependent. Such a material is typically a crystal having an index of refraction along one direction that is different (higher) than that along the two orthogonal directions. This direction is referred to as the "extraordinary" direction (as opposed to "ordinary" for the other two directions). The plate is cut perpendicular to one of the ordinary axes, and the thickness is cut to allow the difference in electrical pathlength for two orthogonal polarizations to be π or $\pi/2$ (for a half-wave or quarter-wave plate, respectively) at a chosen wavelength. The *birefringence* of a material is defined as

$$\Delta n \equiv n_{\mathrm{e}} - n_{\mathrm{o}}, \tag{4.13}$$

where n_e is the index along the *extraordinary* direction, and n_{o} is the index along the *ordinary* direction. If $n_{\mathrm{e}} < n_{\mathrm{o}}$ ($n_{\mathrm{e}} > n_{\mathrm{o}}$), the extraordinary (ordinary) and ordinary (extraordinary) directions in the interface plane are called the "fast" and "slow" axes, respectively.

Quartz and sapphire are commonly used for such applications, since they are both birefringent. Goldsmith (1998) provides a thorough summary of the optical characteristics of both of these materials. Quartz is transparent both in the near-infrared and in the far-infrared through millimeter part of the spectrum. It has a birefringence, $\Delta n_{\mathrm{quartz}} \sim 0.05$. Sapphire on the other hand has a significantly higher birefringence, $\Delta n_{\mathrm{sapphire}} \sim 0.3$. Because of its relatively higher birefringence, sapphire wave plates can be made thinner relative to quartz wave plates. Sapphire has a slightly higher index and loss tangent than quartz, so that the optical advantage of thinner wave plates from insertion and ohmic loss perspectives is a trade that must be considered.

Birefringence can also be synthesized via a metamaterial approach. By machining sub-wavelength features into a homogeneous material as a simulated dielectric, it is possible to generate an effective birefringence. For example, this approach has been implemented for astrophysical polarimeters using rexolite by Glenn et al. (1997). More recently, silicon wave plates have been developed by sawing anisotropic features in a silicon wafer (Coughlin 2018).

We remind the reader that we are again operating in the paraxial limit, and hence, we assume that the wave plate is operating such that the rays of light all propagate parallel to the optical axis. For illustrative purposes, we also ignore any reflections at the front and back surfaces of the plate. In this simple picture, the operation of the wave plate on the incoming radiation can be modeled by a relative phase delay, δ, that is given by

$$\delta = 2\pi(n_e - n_o)\frac{t}{\lambda}, \tag{4.14}$$

where n_e and n_0 are the index of refraction for the extraordinary and ordinary rays, respectively. The wavelength is given by λ and plate thickness by t. For the case of a half-wave plate, the thickness is selected such that $\delta = (2N - 1)\pi$, corresponding to the half-wave condition for a particular design wavelength. Here, N is the effective delay order. It is desirable to keep N as low as practically possible. The plate is rotated about an axis perpendicular to the principle axis of the optical system.

Before concluding the topic of birefringent plates, it should be noted that in the previous discussion, we have ignored the reflections at the interfaces between the plate and the rest of the optical system. These interfaces present a mismatch in dielectric constant (or wave impedance from a guided wave perspective) and result in reflections at the surfaces. These cause both a loss in efficiency in addition to frequency-dependent resonances within the plate. For this reason, at these wavelengths, antireflective coatings are employed for dielectric wave plates. These consist of a quarter wave layer of material with an index that is the geometric mean between the plate's index and that of the ambient medium (usually a vacuum). This ideally allows perfect cancellation of reflected waves at a single wavelength/frequency, but the efficacy of the coating degrades away from the design frequency. The larger birefringence and higher index of refraction of sapphire ($\sim$3.0) relative to quartz ($\sim$2.1), means that total and polarization-dependent reflections can be more of an issue for the former and may require more effort in anti-reflection coating.

4.4.4.2 Other Wave Plate Implementations

A related phase-introduction technique can be used if the wave plate is to be used in reflection. For this implementation, a wire grid suspended above and parallel to a mirror can be used to introduce a path length difference between orthogonal linear polarizations. Figure 4.9 shows this situation. The polarization parallel to the grid wires is reflected off of the front surface of the grid; the orthogonal polarization is transmitted by the grid and reflects off the mirror. The path difference between the

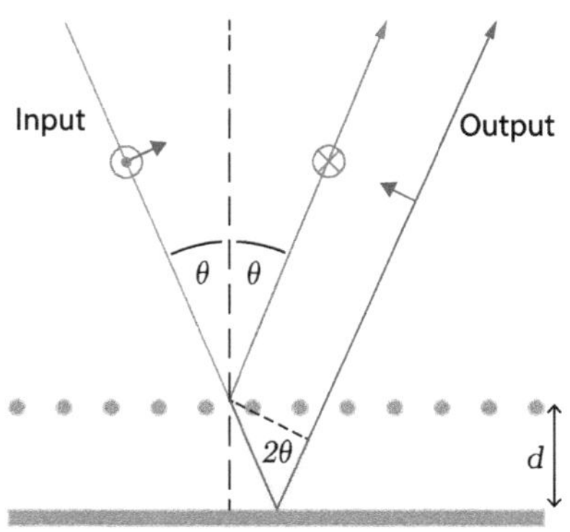

Figure 4.9. A grid and a mirror can be used to introduce a phase delay between two orthogonal polarizations. The physical path difference introduced by this device is highlighted in yellow.

two polarization is shown in yellow in the figure. For a plate separation, d, the phase delay between the two polarizations is

$$\phi = 2\pi \frac{d}{\lambda} \cos \theta, \tag{4.15}$$

where λ is the wavelength in question. The mirror-plate distance, d can be adjusted to set the phase delay at λ to be π, for a half-wave plate and $\pi/2$ for a quarter-wave plate. Such a modulator has been used on the PolKa polarimeter on APEX (Siringo et al. 2004) for the LABOCA camera operating at 870 μm as well as for the Cosmic Foreground Explorer (COFE; Leonardi et al. 2006) and CLASS (Shi et al. 2024). Because there are no dielectrics involved, this architecture has extremely low loss and does not require antireflective coatings. Potential considerations include the fact that this wave plate architecture is by nature used at non-normal incidence and the two beams are laterally displaced upon exiting the modulator. In addition, the phase delay introduced will vary across the beam more severely than for a wave plate implemented for a paraxial beam.

4.4.4.3 Finite Bandwidth Considerations

Both realizations described above provide excellent half-wave plate performance at a single wavelength/frequency but degrade as the operating bandwidth of the instrument increases. It is difficult to define a unique modulation efficiency for a device, since the degree to which a wave plate approximates its ideal behavior is inextricably coupled to its usage in an instrument. Ultimately, the requirements regarding wave plate performance is dependent on the particular application and is often dependent on factors such as the frequency spectrum of the source. In light of this, we will focus on describing the physics as a guide to considerations for implementations of a wave plate for an end goal. Often some sort of metric for polarization efficiency and systematic uncertainty can be estimated using a model for the particular system of interest. In this way, the impact of non-ideal behavior can be assessed for a particular application. To do this, we utilize the Poincaré sphere as an analysis tool.

As previously stated, a phase delay between two orthogonal polarizations can be represented on the Poincaré sphere as a rotation about the diameter connecting the two orthogonal polarization states. The magnitude of the rotation is equal to the phase delay. For a half-wave plate, the target phase delay is π; however, for the architectures discussed above, this ideal phase delay can only be exactly achieved at a single wavelength. A finite bandwidth leads to dispersion, which is dependent on the angle of the fast/slow axes orientation relative to the input polarization state. As an example, Figure 4.10 shows the dispersion for a half-wave plate with 20% bandwidth. For illustration purposes, we have assumed a horizontally polarized input polarization ($+Q$) and show the dispersion for relative orientation of the wave plate axis and polarization direction of 0°, 22.5°, 45°, and 67.5°.

Such single layer wave plates have been used for instruments having $\Delta\nu/\nu$ up to 0.3, with decreasing polarimetric efficiency as this increases. It is useful to note that

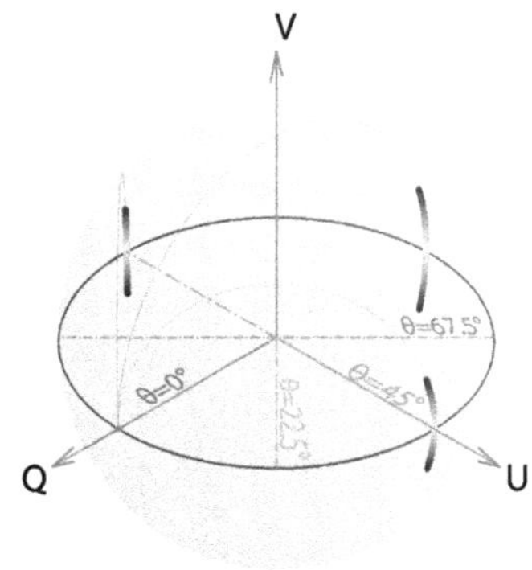

Figure 4.10. The dispersion for the action of a half-wave plate is shown. The input polarization is in the $+Q$ orientation, and the dispersion is illustrated over a 20% passband for four orientations of the fast/slow axis of the wave plate relative to the input polarization angle. Both the efficiency (dispersion along out of the Q–U plane) and the output angle (dispersion in the Q–U plane) are dependent on the orientation of the wave plate and the spectral content of the signal of interest.

the effect of the dispersion changes with the relative angle of the input polarization and the fast/slow wave plate axis. When this angle is either 0° or 45°, the dispersion lies along a great circle in Poincaré space. In these case the final polarization states are mixed between desired polarization and Stokes V, For 0°, the dispersion is zero; for 45°, the dispersion is maximal. For angles other than these, the dispersion is less than the maximum, but also has a component that is dispersed along the Q–U plane in addition to the dispersion in the $\pm V$ direction. The consequence of this is that the effective rotation angle of the wave plate is dependent on the spectral content of the input signal. This is likely to be a small effect in most applications, but should be considered in broadband situations.

4.4.4.4 Material Considerations

Polarimeters operating in the far-infrared typically have fractional bandwidths $\Delta\lambda/\lambda \sim 0.1$–$0.2$, and thus commonly employ single crystal wave plates. Because the wavelengths are short, and thus wave plates with small birefringence can be made at reasonable sizes, crystalline quartz is a reasonable material at far-infrared and submillimeter wavelengths. Early FIR polarimeters (Novak et al. 1989a; Platt et al. 1991) operating on the Kuiper Airborne Observatory were among the first to utilize quartz wave plates at these wavelengths. Since then, many others have utilized quartz wave plates (Kusaka et al. 2014a).

Operation at longer wavelengths requires thicker wave plates for a given material. In addition, apertures for CMB studies tend to be large to enable a high throughput. Available crystalline quartz is limited to aperature diameters significantly smaller than desirable to support the focal plane array optics. Thus, polarimeters at millimeter wavelengths are driven toward the use of synthetic sapphire half-wave plates. These include the Atacama B-Mode Search (Kusaka et al. 2014a; EBEX Collaboration et al. 2018), and Spider (Bryan et al. 2010), among others.

4.4.4.5 Pancharatnam Wave Plates

Pancharatnam (1955) realized that a stack of wave plates could be used to synthesize a composite wave plate with a significantly broader bandwidth modulation efficiency than a single-layer plate. In the original work, this was applied to a circular polarizer (quarter-wave plate). In this case, it was shown that two birefringent plates can be used to compensate for the frequencies within the band for which the polarization delay was greater than or less than $\pi/2$.

This concept was later extended to the FIR through the millimeter part of the spectrum (Savini et al. 2006; Pisano et al. 2006) for the QuAD cosmic microwave background polarimeter (Hinderks et al. 2009) using a five-plate stack. Such achromatic stacks of wave plates have been used in other instruments as well, including BLASTpol (Fissel et al. 2010), Spider (Filippini et al. 2010; Bryan et al. 2010), Pol-2 on SCUBA-2 (Friberg et al. 2016), and TOLTEC (Lee et al. 2022). Recently, various achromatic wave plate designs have been considered for LiteBIRD (Komatsu et al. 2021).

Figure 4.11 shows an example of a three-plate combination from this work that illustrates the technique for compensating the polarization rotation for frequencies across a wide band. In this case, the bandwidth under consideration is 20%. The initial polarization state is taken to be a pure $-U$ state for all frequencies within the band. The first wave plate is oriented at an angle of 58.5° from the fast axis of the wave plate. This operation is shown by the rotation marked ①. The next wave plate is oriented at an angle of 0° with respect to the same axis. This rotation is shown as ②. At this point, the dispersion across the band is quite large. The final wave plate is aligned parallel to the first. This rotation, designated by ③, brings the wavelengths across the band into the final position close to the Q–U plane. In the

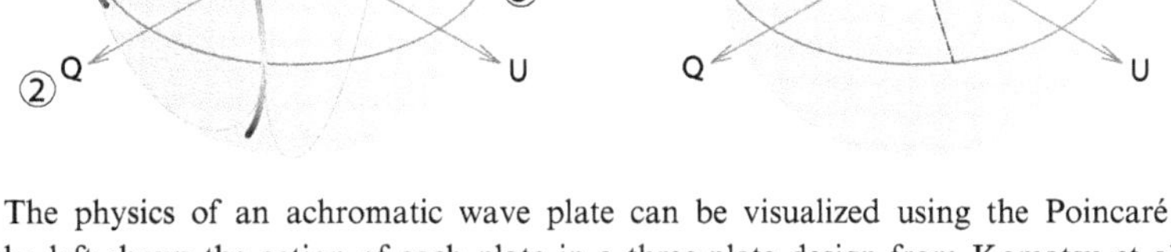

Figure 4.11. The physics of an achromatic wave plate can be visualized using the Poincaré sphere. The diagram on the left shows the action of each plate in a three-plate design from Komatsu et al. (2021). An example initial polarization state ($-U$) is shown by a gray point. The first rotation (1) is about the axis oriented 58.5° with respect to the horizontal ($+Q$). The dispersion of the initial polarization state for a 40% bandwidth is shown. The next plate (2) further disperses the band over the surface of the sphere. The final plate (3), which has the same orientation of the first, then compensates for this dispersion. On the right, the effective wave plate angle is shown along with the dispersion of the final polarization state from the dielectric stack. The dispersion in this case is confined mostly to the Q–U plane. The effective angle of the wave plate is dependent on the frequency content of the input signal.

right hand side of the figure, the initial state and set of final states are shown along with the effective axis of rotation.

The Pancharatnam design increases the efficiency of a wave plate across a larger bandwidth than a single plate by concentrating the frequencies within the band within a region close to the Q–U plane in the final state. The trade is that there is an increased dispersion along the direction of the Q–U plane. This means that the output polarization angle is frequency-dependent, which must be considered in the context of the desired implementation. Often, such plates are used for a single focal plane with multichroic channels (i.e., several bandpasses in a single optical system). As a result, the effective wave plate angle needs to be calibrated for each frequency channel in the system.

4.4.4.6 Metal Mesh Technologies

A second means to achieve broadband performance for wave plates is via dielectrically-embedded metal mesh technology. The principle of operation here is that anisotropic metamaterial layers are superposed: one set presents an inductive layer for one linear polarization; the other presents a capacitive layer for the orthogonal polarization. These have the effect of introducing phase delays in opposite directions for the two orthogonal linear polarizations. A total differential phase delay of π can be achieved via a metamaterial device (Pisano et al. 2012; 2016). Quarter-wave plates can be produced by setting the phase delay to $\pi/2$. The bandwidth of such a device can be increased by adding multiple layers for each of the capacitive and inductive functions. For example, the NIKA-2 wave plate design utilizes six layers for each of the two polarizations.

Recently (Pisano et al. 2022) have implemented extremely large bandwidth half-wave plates by via artificial dielectrics. In this case, the two orthogonal linear polarizations are presented with electric and magnetic boundaries, respectively, providing a broadband differential phase shift of π between the polarizations.

4.4.5 Variable-Delay Polarization Modulators

Variable-delay polarization modulators (VPMs) fix the direction of the phase delay introduction and then vary the magnitude of the phase. The Poincaré sphere representation for a VPM is shown in Figure 4.12. The phase delay is introduced in the same direction as the dispersion, and the dispersion increases in the direction of the phase delay. This is identical to the action of a Martin–Puplett interferometer (Martin & Puplett 1970). Thus, the Fourier transform of the interferogram is the spectrum of the linear (or circular) Stokes parameter (Chuss et al. 2006).

As mentioned above, the Martin–Puplett interferometer performs exactly this function, and in principle one could be used for a polarization modulator. However, a more compact architecture, such as that shown in Figure 4.9 is common. Here, a polarizing grid is positioned parallel to and in front of a reflecting surface. The separation between the mirror and grid is carefully varied. As for the case of the half-wave plate, it is critical to track the state of the system for later demodulation and detection.

With the architecture described, the zero phase delay point is not practically attainable, as it would require collision of the grid and mirror. A typical use for the

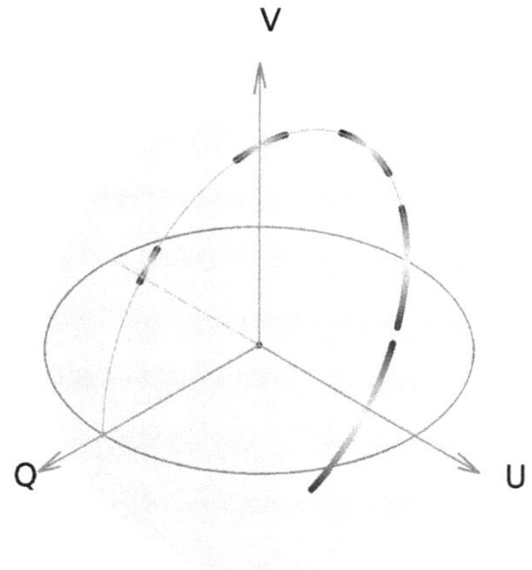

Figure 4.12. Modulation path on the Poincaré sphere for a variable-delay polarization modulator (VPM). In this case, the modulation direction and the dispersion direction are co-aligned. In addition, the dispersion increases with the phase delay. This is identical to the action of a Martin–Puplett interferometer.

VPM is to set the device between phase delays corresponding to integer multiples of π. Usually these correspond to π and 2π to minimize dispersion. The state of the system is varied between these, and the modulator acts as a switch.

VPMs were first implemented in the Hertz/VPM experiment (Krejny et al. 2008) as a proof-of-concept in a "dual switching" configuration in which two VPMs were used—one to select between Q and U and the other to modulate the polarization by 90°. Chuss et al. (2012a) investigated non-ideal effects contributing variable emission from the device. VPMs were constructed for the PIPER balloon-borne CMB experiment (Chuss et al. 2010; Gandilo et al. 2016) These VPMs were designed for cryogenic operation (Chuss et al. 2014). The CLASS CMB experiment (Essinger-Hileman et al. 2016) has implemented VPMs. These have been characterized by Harrington et al. (2021), and produced initial measurements of 70% of the microwave sky at 40 GHz (Eimer et al. 2024). The VPM's residual sensitivity to circular polarization has enabled CLASS to characterize the circular polarization of the atmosphere at 40 GHz (Petroff et al. 2020) and has produced upper limits on cosmic circular polarization (Padilla et al. 2020; see Chapter 7).

4.5 Incoherent Detection

In the far-infrared through millimeter range of the spectrum, detector development is challenging. Unlike the optical and near-infrared, where industrial demand and availability of band-gap materials has driven development of megapixel focal planes, far-IR detector technology development has relied on astronomical research efforts. However, in recent decades, there has been overlap with the quantum computer industry in terms of cryogenic systems and superconducting circuit elements. In this section, we give a brief introduction to selective technologies. To learn more see Richards (1994); Zmuidzinas & Richards (2004).

One key concept that is crucial to maximizing sensitivity at these wavelengths is for the noise budget to be dominated by photon noise from the scene. Ideally, this is from the source itself, but for non space-based telescopes, the *background limit* is often set by the atmosphere, ground pick-up, and instrument emission. In practice, the detectors and their relevant readout electronics must contribute significantly less noise than the source. Noise in detectors is dependent on temperature, so at these wavelengths, background limited detectors must be cooled, often to significantly below 1 K. In addition, the $1/f$ spectrum of the detector must also be subdominant or otherwise mitigated. We offer a brief review of several relevant technologies here. Later in this chapter, we give some examples of instruments in which these are implemented.

4.5.1 Bolometers

Thermal bolometric detectors directly convert absorbed radiation into heat that creates a physical signature that can be sensed. Typically, they contain a thermistor element that has a resistance that is strongly dependent on temperature. The change in resistance is read out via an appropriate electric circuit. The basic thermal circuit for a bolometer is shown in Figure 4.13.

The key physical process is the coupling of light via an absorber into heat within the bolometer itself, which has a heat capacity, C. Heat then flows via a thermal link characterized by thermal conductance, G to a thermal bath. The bolometer noise is given by Low (1961); Mather (1982)

$$NEP \simeq \sqrt{4kT^2G}, \tag{4.16}$$

where k is Boltzmann's constant and T is the operating temperature of the sensor.

For a superconducting transition edge sensor (TES), the device is operated on transition, which is designated as the critical temperature, T_c, of the superconducting material. From the device structure the impedance is typically low and readout by a quantum-limited amplifier. The thermal time constant is $\tau = C/G$. In operating bolometers, the response time is typically slightly lower than this due to electro-thermal feedback (Low 1961; Irwin & Hilton 2005).

Semiconducting bolometers are operated in a current-biased mode in which a load resistor is placed in series with the bolometer. Typically these thermistors are doped semiconductors. Early single bolometers were gallium-doped germanium and

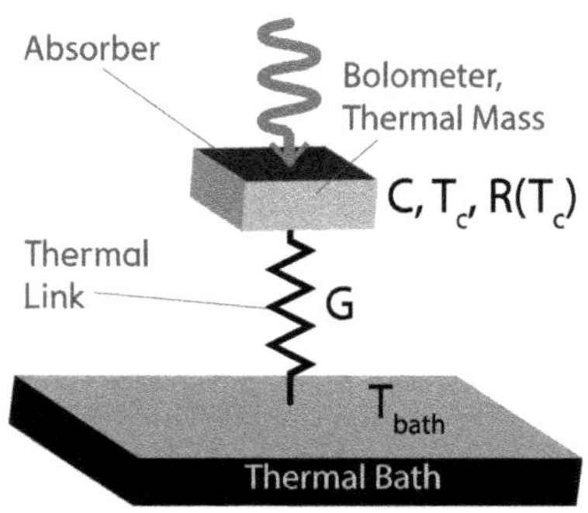

Figure 4.13. The thermal circuit for bolometers is shown. Detector noise is dependent on the value of the thermal conductance (G), and the time constant for the device is $\tau = C/G$.

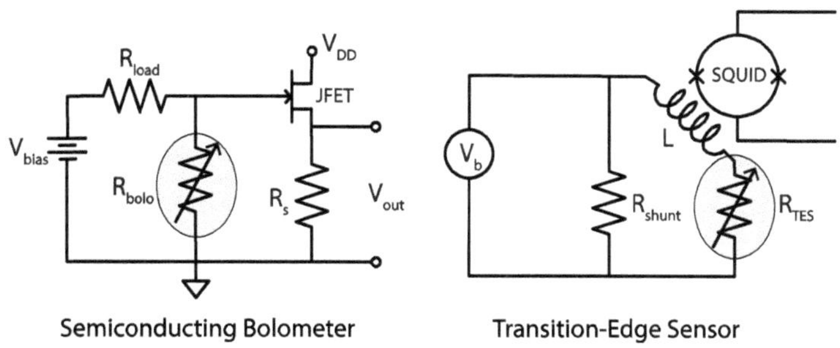

Figure 4.14. Left: The readout circuit for a semiconducting bolometer. The bolometer is current-biased through the use of a series load resistor, R_{load}. The signal is read out as a voltage across the bolometer. A JFET source-follower transforms between high impedance and low impedance. Right: The readout for a superconducting transition-edge sensing (TES) bolometer is shown. The bolometer is voltage-biased so that it is stable on its superconducting transition. The absorption of photons increases the resistance of the TES and therefore reduces the current through the sensing element. The current is read out by a superconducting quantum interference device (SQUID).

arsenic-doped Silicon. More recently, neutron-transmutation devices (NTD; Haller et al., 1994) have been implemented. See Figure 4.14 (Left). As energy is absorbed in the bolometer, the temperature and resistance rise, resulting in a larger voltage. This resulting signal has a high impedance, and thus can be subject to microphonic pickup (i.e., noise induced through vibrational modulation of the readout circuit). A JFET-based source-follower is included in the readout circuitry to convert the signal to a lower-impedance. Semiconducting bolometers require a single channel for each detector.

Superconducting transition-edge sensor (TES) technology provides a steep $R(T)$ relation when operating on the devices' superconducting transition. This offers excellent sensitivity, but because the function is so steep, current-biasing results in instability. Instead, TESes are voltage biased using a shunt resistor (Irwin & Hilton 2005; see Figure 4.14 (right)). In this configuration, the total power dissipated in the TES is constant, resulting from a sum of the absorbed optical power and the Ohmic dissipation from the bias circuit. As the optical load increases, the resistance of the TES also increases, resulting in a decrease in current and corresponding Ohmic dissipation. This keeps the TES biased on its superconducting transition. The current in the bias circuit is then an inverse proxy for the optical power absorbed. The current changes in the bias circuit are read by a superconducting quantum interference device (SQUID). Because the signal is low-impedance, microphonic pickup tends to present a lower detector system implementation challenge (i.e., than commonly encountered in high-impedance cryogenic systems). However, a major concern with TES circuits is the presence of magnetic fields, which can both alter the superconducting properties of the TES itself and interfere with the SQUID.

Independent of the technology of the sensor, once the background limit is reached, the only means for increasing mapping speed (in both total power imaging and in polarimetry) is to increase the number of independent resolution elements in the focal plane. Thus, development of large arrays of detectors have been a priority for the past few decades. This has motivated both time-domain (TDM; Irwin 2002) and frequency-domain multiplexing (FDM; Lanting et al. 2005, respectively) strategies. These advances have enabled ~kilopixel superconductive detector arrays.

4.5.2 MKIDs

Microwave kinetic inductance detector (MKID; Day et al. 2003) technology provides improvements over TESs in both ease of fabrication and in multiplexing factor. In this case, a coaxial line can read up to ~1000 detectors with microwave readout electronics. The basic detection element is a thin-film metallic superconductor that is placed in a microwave resonant circuit (Zmuidzinas 2012). Incoming photons above the gap energy break Cooper electron pairs in the thin film, thereby altering the inductance of the superconductor. This changes the center frequency and quality factor of the resonant circuit. The induced frequency shift is then detected by the readout. Each detector is assigned a unique resonant frequency, so many such detectors can be read out through a coaxial line via superposition of these resonances.

4.6 Signal and Noise

In the far-infrared through millimeter part of the electromagnetic spectrum, the scene is complicated by radiation that enters the instrument that is not associated with the target source. These can be astronomical in the case of observations of the cosmic microwave background and can be partially polarized. Foregrounds can come from atmospheric emission for non-space-based observatories and can also result from emission from the telescope itself. Often the undesired power is much larger than that of the source to be measured. In such cases, the brightness of the foregrounds needs to be taken into account when anticipating the loading on the detectors and ultimately sets the dynamic range of the detectors. As an example, for ground-based polarimetry at 350 μm, the polarized astrophysical signal from a typical molecular cloud can be a factor of $\sim 10^6$ below the total power on the detectors (Chuss et al. 2004). In the case of the cosmic microwave background, the well-measured E-mode polarization is $\sim 10^{-7}$ of the unpolarized CMB. For dust emission, the unpolarized part of the astrophysical signal is typically 5–100 times larger than the polarized component.

To complicate things further, some of this emission, namely that from the atmosphere and instrument can be *non-ergodic* in that it typically has a power spectral density with a functional form commonly approximated by $\propto 1/f^{\alpha}$. As will be shown from the Wiener–Khinchin theorem, such a signal is correlated in the time domain. Fundamentally, this component of the signal exhibits variations on time scales larger than the total integration time, and thus a mean cannot be uniquely defined. Noise having a similar spectral profile can also originate within the instrument. Much of the challenge of observing at these wavelengths centers around mitigating this component of the noise through synchronous detection strategies and calibration to limit the variance on long time scales.

4.6.1 The Nyquist–Shannon Sampling Theorem

This is not a book on signal processing; however, in the acquisition, reduction, and analysis of astronomical polarimetry data, an elementary understanding of signal processing is essential. Specifically, thinking in terms of the Fourier domain can

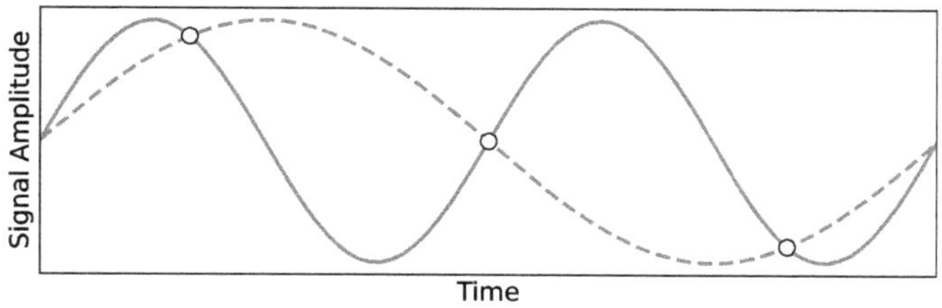

Figure 4.15. Aliasing can occur when the sample rate of data (yellow dots) is too sparse, leading to misidentification of a real high frequency signal (red solid line) as a lower frequency signal (dashed blue line).

often lead to understanding inaccessible in the complementary space or time domain. Fundamental to any endeavor in Fourier analysis is the Nyquist–Shannon Theorem. This sets the theoretical lower limit for sampling rate of a continuous function without information loss and provides a key insight into the link between measurement bandwidth and integration time.

One encounters the idea of a minimum sampling rate in everyday life. For example, one can observe a fan or airplane propeller as it is spun up. The blades appear to go faster and faster up to a point. Then, at some point, the fan appears to rotate backwards. The sampling rate of the observer's eye–brain system is too slow to recognize the faster angular frequency, so it identifies it as a lower frequency. This is called *aliasing* and is graphically shown in Figure 4.15. In this situation, the time-sampled data points are shown as yellow dots. The true signal is the red curve, but the sample rate is too sparse, and the signal is identified as the lower frequency dashed blue curve.

For additional insight we include a derivation of the Nyquist–Shannon sampling theorem. The upshot is the well-known result that in order to discretely sample a function, one needs to sample at a frequency of at least twice that of the maximum frequency at which the signal contains power. We follow Shannon's approach (Shannon 1949), which shows this explicitly for temporal signals, but we remind the reader that this concept is equally applicable to spatial patterns as well (or any pairs of variables related by a Fourier transform). Quoting from Shannon (1949), the Nyquist–Shannon sampling theorem is as follows.[1]

The Nyquist–Shannon Sampling Theorem

If a function $f(t)$ contains no frequencies higher than ν_c cps [Hz], it is completely described by giving its ordinates at a series of points spaced $1/2\nu_c$ seconds apart.

[1] In Shannon's paper, he uses W for the frequency cutoff of the spectrum. We opt for a more intuitive ν_c for this quantity. In addition, we define the Fourier transform in terms of the frequency, $\nu = \omega/2\pi$, rather than ω so that we can compare it directly with the sampling rate.

The function $f(t)$ can be written in terms of its Fourier transform, with its spectrum given by $F(\nu)$,

$$f(t) = \int_{-\infty}^{\infty} F(\nu)e^{i2\pi\nu t} d\nu. \tag{4.17}$$

A sample spectrum, $F(\nu)$ is shown on the left side in Figure 4.16A. If $F(\nu)$ has no support outside of the frequency interval, $[-\nu_c, \nu_c]$, the limits of integration for the angular frequency, ν, can be set to $[-\nu_c, \nu_c]$,

$$f(t) = \int_{-\nu_c}^{\nu_c} F(\nu)e^{i2\pi\nu t} d\nu. \tag{4.18}$$

The function $f(t)$ is shown in Figure 4.16A on the right.

We can address the problem of determining the sampling by treating a similar function, $\tilde{F}(\nu)$ that has the same functional form within the $[-\nu_c, \nu_c]$ range, but repeats with a periodicity $2\nu_c$. This is illustrated in Figure 4.16B. Because of its periodicity, $\tilde{F}(\nu)$ can be described by a Fourier *series* rather than a Fourier *transform*,

$$\tilde{F}(\nu) = \sum_{n=0}^{\infty} \tilde{f}(t_n)e^{i2\pi\nu t_n} \tag{4.19}$$

There is a discrete set of sinusoidal functions, $e^{i2\pi\nu t_n}$, that span the space of all periodic functions with periodicity 2ν (in the frequency domain), and these are of course those that have an integer number of oscillations within $[-\nu_c, \nu_c]$. We introduce the notation $\tilde{f}(t_n)$ as the weights for the linear combination that comprises the spectrum. Because both $F(\nu)$ and $\tilde{F}(\nu)$ have the same form within $[-\nu_c, \nu_c]$, $\tilde{f}(t_n)$ is a discrete function that has the same values as $f(t)$ at the discrete points, $t_n = n/2\nu_c$ (see Figure 4.16B, right, for a representation of $\tilde{f}(t_n)$),

$$f\left(\frac{n}{2\nu_c}\right) = \tilde{f}\left(\frac{n}{2\nu_c}\right) = \int_{-\nu_c}^{\nu_c} \tilde{F}(\nu)\exp\left(2\pi i\nu\frac{n}{2\nu_c}\right)d\nu. \tag{4.20}$$

As we have stated, this quantity gives the weights of the basis functions that describe $\tilde{F}(\nu)$ over all frequencies, or $F(\nu)$ over frequencies within $[-\nu_c, \nu_c]$. Thus, we can also interpret this as values of the function $f(t)$ that have been discretely sampled at $\Delta t = n/2\nu_c$. In other words, the Fourier coefficients in the time domain that represent the expansion of the spectrum, $\tilde{F}(\nu)$, are the time-domain points at which the waveform is required to be sampled for complete reconstruction of the frequencies in the signal.

Finally, we return to the goal of reconstructing $f(t)$ completely. To do this, we will start in the frequency domain and then move to the time domain. To recover the original spectrum $F(\nu)$ from $\tilde{F}(\nu)$, we can apply a unity amplitude boxcar filter, $H(\nu)$, to $\tilde{F}(\nu)$,

$$F(\nu) = \tilde{F}(\nu) \times H(\nu), \tag{4.21}$$

where,

$$H(\omega) = \begin{cases} 1 & \text{if } \nu \in [-\nu_c, \nu_c] \\ 0 & \text{otherwise} \end{cases}. \tag{4.22}$$

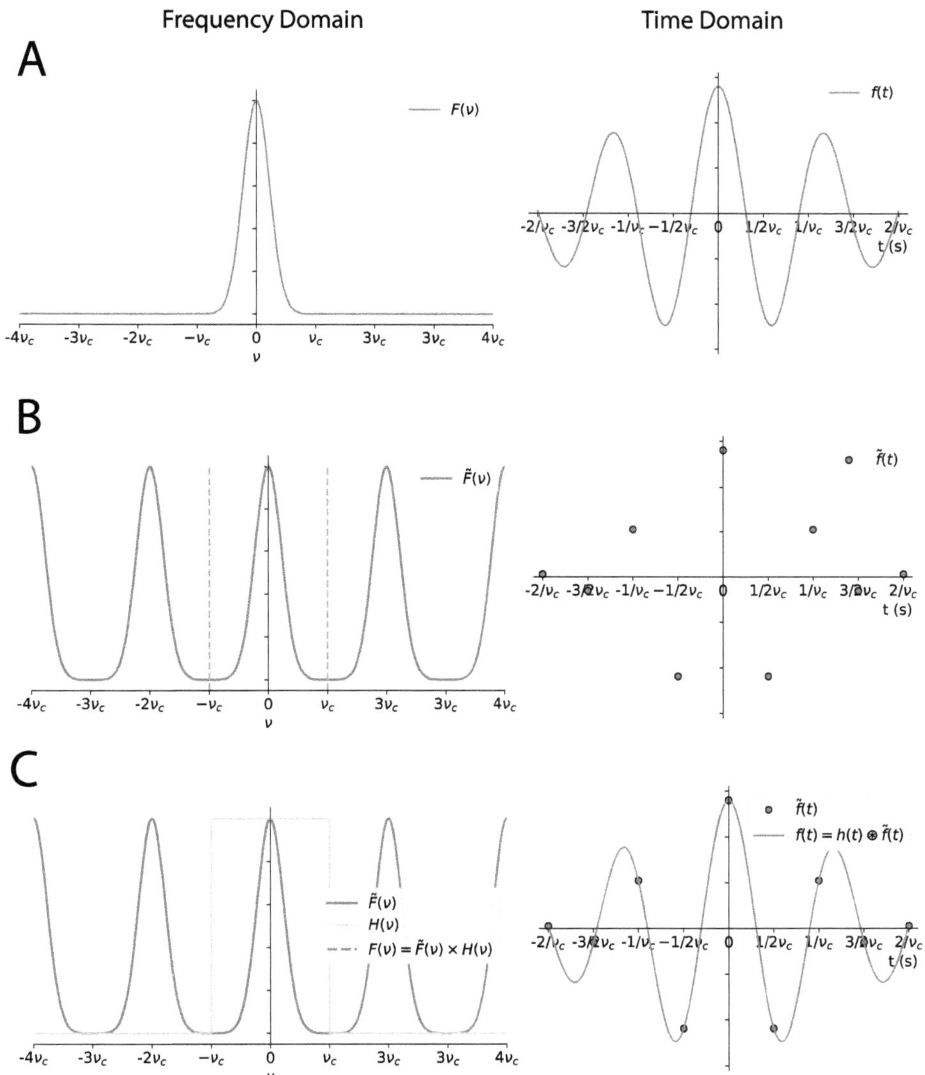

Figure 4.16. A sample spectrum of a signal, $F(\nu)$, whose support is confined to the angular frequency interval $[\nu_c, \nu_c]$ is shown (A) along with a corresponding time stream taken over an interval of $[-2/\nu_c, 2/\nu_c]$. To determine the minimum sample rate for this timestream without loss of information, we imagine a periodic function which is equal to our initial $F(\nu)$ over the relevant time interval (B). This periodic function, $\tilde{F}(\nu)$ has a discrete Fourier transform. The original spectrum can be recovered from the new function by multiplying by a boxcar function, $H(\nu)$, that has support over the defined interval (C). In the time domain, this is the convolution of the Fourier transforms of the two functions, which recovers the original timestream fully.

This is schematically represented in the left panel of Figure 4.16C. The Fourier transform of the boxcar function is the sinc function, $h(t) = \text{sinc}(2\pi\nu_c t)$. The convolution theorem states the Fourier transform of the product of two functions is the convolution ($\circledast$) of their Fourier transforms. Therefore,

$$f(t) = \tilde{f}(t) \circledast \text{sinc}(2\pi\nu_c t) = \int_{-\infty}^{\infty} \tilde{f}(\text{t}')\text{sinc}[2\pi\nu_c(t - \text{t}')]d\text{t}'. \qquad (4.23)$$

In the conjugate time domain, this corresponds to the convolution of the function $\text{sinc}(2\pi\nu_c \text{t})$ with the discrete samples, $f(n/2\nu_c)$. Here, we have defined $\text{sinc}(x) = \frac{\sin x}{x}$.

The reconstructed function $f(t)$ is equal to the original and shown superposed on the sampled points in Figure 4.16(C) (right). Thus, we have shown that the sampling needed to completely reconstruct a signal confined to $[-\nu_c, \nu_c]$ is $\Delta\text{t} = 1/2\nu_c$ or $\nu_{\text{Nyquist}} = 2\nu_c$, in agreement with the Nyquist–Shannon sampling theorem.

4.6.2 Integration Time and Output Bandwidth

An interesting consequence of the Nyquist–Shannon theorem relates to the relationship between integration time, τ, and measurement bandwidth, $\Delta\nu$. Intuition suggests that for higher τ, $F(\nu)$ should be constrained with increasing fidelity. Returning to the example above where time samples correspond to the amplitudes of the frequency mode, additional samples correspond to the amplitudes of modes that oscillate more in the frequency domain. This increases the resolution of the passband, which we identify with *measurement bandwidth* (sometimes called *video bandwidth*), $\Delta\nu$.

Using Shannon's argument above dictates that a sampling time interval and a frequency bandwidth are related via the Nyquist condition $\Delta t \to 1/(2\nu)$. The same argument can be applied if one reverses the roles of the domains such that for an integration time, we get an equivalent output bandwidth, $\Delta\nu \to 1/(2\tau)$. This equivalence (or conversion) is sometimes encountered in calculating the sensitivity of a polarimetric system. Specifically, one encounters this when using a sensitivity metric for an instrument or detector to calculate the explicit uncertainty in a measurement.

The *Noise Equivalent Power NEP* of a system is the incident power (on a detector) that produces a unity signal-to-noise ratio in a specified time interval. NEP is perhaps most commonly measured in units of power per square root bandwidth, for example, W $\text{Hz}^{-1/2}$. This corresponds to the power, in Watts, detectable at unity signal-to-noise in an output bandwidth of 1 Hz. From the above argument, a 1 Hz output bandwidth corresponds to an integration time of 0.5 seconds.

Because the level of sensitivity reached for an observation depends on the square root of the observation time, it is sometimes useful to express the NEP in terms of power square root time (e.g., W $\text{s}^{1/2}$). In this case the NEP value refers to the power detectable for a unity signal-to-noise ratio in a 1 second integration. Thus, for the same detector, using the second set of units results in a lower value for the NEP by a factor of $\sqrt{2}$ because the integration time is implicitly larger by a factor of 2. For example,

$$\text{NEP(units of W Hz}^{-1/2}) = \sqrt{2}\ \text{NEP(units of W s}^{1/2}).$$

This same argument applies to other Noise Equivalent Units: *Noise Equivalent Flux Density (NEFD)* or for CMB studies, *Noise Equivalent Temperature (NET)* or *Noise Equivalent (Stokes) Q (NEQ)*.

4.6.3 The Convolution Theorem

In experimental settings an appreciation of the relationships between the time (or space) and frequency (or angular) domains can provide extremely useful insights. Complete descriptions of statistical optics, signal analysis, and data reduction are beyond the scope of this book, but we do mention two Fourier techniques that are powerful, namely the convolution and Wiener–Khinchin theorems, which are closely-related.

The Convolution Theorem

The convolution of two functions, $\{U \circledast V\}(\tau)$, is equal to the inverse Fourier transform of the product of the Fourier transform of the two functions.

Proof:

$$\begin{aligned}
\{U \circledast V\}(\tau) &= \frac{1}{\sqrt{2\pi}} \int_{-\infty}^{\infty} U^*(t)V(t-\tau)dt \\
&= \frac{1}{(2\pi)^{3/2}} \int_{-\infty}^{\infty} \left[\int_{-\infty}^{\infty} \tilde{U}^*(\omega)e^{i\omega t}d\omega\right]\left[\int_{-\infty}^{\infty} \tilde{V}(\omega')e^{-i\omega'(t-\tau)}d\omega'\right]dt \\
&= \frac{1}{(2\pi)^{3/2}} \int_{-\infty}^{\infty} d\omega\ \tilde{U}^*(\omega) \int_{-\infty}^{\infty} d\omega' \tilde{V}(\omega')e^{i\omega'\tau} \int_{-\infty}^{\infty} dt\ e^{i(\omega-\omega')t} \\
&= \frac{1}{(2\pi)^{3/2}} \int_{-\infty}^{\infty} d\omega\ \tilde{U}^*(\omega) \int_{-\infty}^{\infty} d\omega' \tilde{V}(\omega')e^{i\omega'\tau}[2\pi\delta(\omega-\omega')] \\
&= \frac{1}{\sqrt{2\pi}} \int_{-\infty}^{\infty} \tilde{U}^*(\omega)\tilde{V}(\omega)e^{i\omega\tau}d\omega \\
\{U \circledast V\}(\tau) &= \mathscr{F}^{-1}(\tilde{U}^*(\omega)\tilde{V}(\omega))
\end{aligned}$$

The convolution theorem comes into play in several contexts over the course of instrument development. One place where this is encountered is in telescope design. In Fraunhoffer diffraction, the electric field distribution at the aperture of a telescope (a pupil) and that at a focal plane are related by a Fourier transform. The field at the pupil is necessarily truncated by the optics and thus can be modeled as the product of the field incident on the telescope and the apodization function. The effect as viewed in the focal plane is the convolution of the *point-spread function (PSF)* with the image that would have been measured with an infinitely-large telescope. In the end, the resultant image recorded is a band limited (or spatially filtered) representation of the input scene.

Another example of where this occurs is in Fourier transform spectro-polarimetry (Martin & Puplett 1970). Here, the spectrum of a source is the Fourier transform of the interferogram. Because the instrument has finite data acquisition time and phase delay limits, the source interferogram is apodized by the instrument. The measured spectrum is the convolution of the source spectrum with the Fourier transform of the apodization function. This is equivalent to the (inverse) Fourier transform of the product of the source interferogram and the apodization function.

Finally, the convolution theorem is useful for considering filtering of time-stream signals. Filtering is a product in the frequency domain that is manifest as a convolution in the time domain. The filter is typically defined by its frequency response (phase and amplitude) and its effect is a multiplication of the frequency-domain filter function with the frequency spectrum of the source. The effect on the time-domain signal from the source is the convolution of the inverse Fourier transform of the filter with the time-domain signal. Note that this is the opposite of apodization, which manifests as a product in the time or space domain and a convolution in the frequency/angle domain.

4.6.4 The Wiener–Khinchin Theorem

The Wiener–Khinchin Theorem

The autocorrelation of a function, $G(\tau)$, is equal to the inverse Fourier transform of the power spectral density (PSD), $\mathscr{F}^{-1}(|U(\omega)|^2)$.

Proof:

$$\begin{aligned}
G(\tau) &= \frac{1}{\sqrt{2\pi}}\int_{-\infty}^{\infty} U^*(t)U(t+\tau)dt \\
&= \frac{1}{(2\pi)^{3/2}}\int_{-\infty}^{\infty}\left[\int_{-\infty}^{\infty}\tilde{U}^*(\omega)e^{i\omega t}d\omega\right]\left[\int_{-\infty}^{\infty}\tilde{U}(\omega')e^{-i\omega'(t+\tau)}d\omega'\right]dt \\
&= \frac{1}{(2\pi)^{3/2}}\int_{-\infty}^{\infty}d\omega\ \tilde{U}^*(\omega)\int_{-\infty}^{\infty}d\omega'\tilde{U}(\omega')e^{i\omega'\tau}\int_{-\infty}^{\infty}dt\ e^{i(\omega-\omega')t} \\
&= \frac{1}{(2\pi)^{3/2}}\int_{-\infty}^{\infty}d\omega\ \tilde{U}^*(\omega)\int_{-\infty}^{\infty}d\omega'\tilde{U}(\omega')e^{i\omega\tau}[2\pi\delta(\omega-\omega')] \\
&= \frac{1}{\sqrt{2\pi}}\int_{-\infty}^{\infty}\tilde{U}^*(\omega)\tilde{U}(\omega)e^{i\omega\tau}d\omega \\
&= \frac{1}{\sqrt{2\pi}}\int_{-\infty}^{\infty}|\tilde{U}(\omega)|^2e^{i\omega\tau}d\omega \\
G(\tau) &= \mathscr{F}^{-1}(|\tilde{U}(\omega)|^2)
\end{aligned}$$

The Wiener–Khinchin theorem is particularly useful when considering noise, for which it reveals the relationship between the characteristics of noise in the time and frequency domains. For example, *white noise* refers to noise with a flat *power spectral density (PSD)*, having equal contributions of power at each frequency.

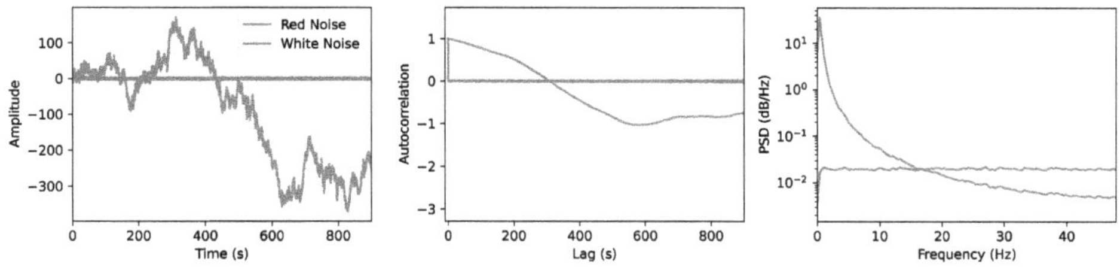

Figure 4.17. An illustration of the Wiener–Khinchin theorem. The time streams, autocorrelations, and power spectral densities for white noise and "red" ($1/f$) noise are compared. Note that for white noise, the PSD is flat and the autocorrelation is given by a δ-function. For 1/f noise, the PSD is formally divergent at zero frequency. Both PSDs have finite spectral energy – in practice – the integral is regularized by the inverse time scale of the observation and the sample frequency, at low and high frequencies respectively.

From the Wiener–Khinchin theorem, one can directly see that such a PSD corresponds to an autocorrelation in the time stream that is a δ-function. The physical interpretation for this is that each sample in the time stream is independent, i.e., the timestream is uncorrelated. The photon noise from a blackbody is a good approximation of this condition.[2]

On the other hand, correlated noise such as $1/f$ noise corresponds to a PSD with structure. For such sources, $G(\tau)$ will explicitly show this correlation on relevant timescales. Correlated noise in polarimetric systems can occur as a result of temperature variations in the foreground emission sources (such as the atmosphere and optics) or as a result of thermalization timescales or recombination noise (e.g., for kinetic inductance detectors) within detectors. This is a major challenge for most data reduction strategies. The difference between white and red ($1/f$) noise in terms of their time streams, autocorrelation functions, and PSDs are shown in Figure 4.17.

An extension to the Wiener–Khinchin theorem for polarimetry was mentioned in Chapter 2. This theorem applies more broadly to each element of the 2×2 mutual coherence matrix. Namely, each element of the matrix is an autocorrelation with zero lag. By the Wiener–Khinchin theorem, an equivalent (frequency-domain) matrix can be used in polarization analysis in which each element of the mutual coherence matrix corresponds to its power spectral density.

4.6.5 Blackbody Radiation

Because blackbody radiation is so central to continuum observations for the wavelengths explored in this book, we review the derivation of the Planck radiation law here. This general derivation follows that covered in numerous statistical mechanics texts (e.g., Reif 1965). The specific energy density (energy per unit volume per unit frequency) of blackbody radiation consists of three broad factors: the energy per state, density of states (states available in a given volume), and the photon occupation number.

[2] There are wavelength-scale correlations in blackbody fields as a result of the nature of propagation (Mehta & Wolf 1964a, 1964b; Mehta et al. 1966; Mehta & Wolf 1967b, 1967a); however, in the time domain, such correlations are much shorter than any practical sampling time, so the approximation that noise from blackbody radiation is uncorrelated is generally justified.

4.6.5.1 Energy Per Photon

The energy per photon is given by the Planck–Einstein relation,

$$E = h\nu. \tag{4.24}$$

4.6.5.2 Density of States

The *density of states* refers to the number of photon states that exist in a given spatial unit. In our case, we are assuming a continuum system in three dimensions, so we care about the states per unit volume. We are assuming the continuum limit in three dimensions here, and polarization will add a degeneracy factor of 2. Similar arguments that parallel this one can be used to develop expressions for thermal noise in bounded systems, for example, Johnson noise and thermal noise in waveguides.

The density of states in three-dimensional space is derived by Reif (1965); Rybicki & Lightman (1979), assuming one has a cubic volume of sides L_x, L_y, and L_z and then assuming in that volume, independent modes of photon energy are those that obey periodic boundary conditions,

$$\psi(\mathbf{r} + \mathbf{L}) = \psi(\mathbf{r}) = \mathrm{e}^{\mathbf{ik}\cdot\mathbf{r}}. \tag{4.25}$$

See Figure 4.18 for an illustration.

The number of states for a given wavevector, $k = 2\pi/\lambda$, in each of the three directions is just the total number of wavelengths that fit within the volume in a given direction. For example, along the x-axis one finds:

$$n_x = \frac{L_x}{\lambda} = \frac{k_x L_x}{2\pi}.$$

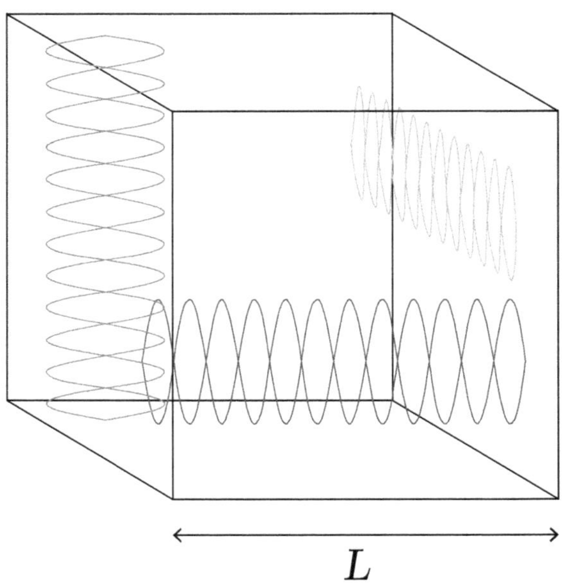

Figure 4.18. Illustration of a set of orthogonally-oriented standing waves in a cubical box with side L.

The number of states within a small wavevector (or frequency) range, dk_x is then,

$$dn_x = \frac{L_x}{2\pi}dk_x.$$

For the full three-dimensional volume this is (Rybicki & Lightman 1979),

$$dN = mdn_x dn_y dn_z = m\frac{L_x L_y L_z}{(2\pi)^3}d^3k = m\frac{V}{(2\pi)^3}d^3k.$$

Here, m is represents the number of polarization states. So, the number density of states per unit volume is,

$$\frac{dN}{V} = m\frac{d^3k}{(2\pi)^3}.$$

Next, we utilize the fact that blackbody radiation is isotropic and write the (k-space) volume element in spherical coordinates,

$$\frac{d^3k}{(2\pi)^3} = m\frac{k^2dkd\Omega}{(2\pi)^3} = m\frac{\nu^2d\nu d\Omega}{c^3}.$$

Here we have used $k = 2\pi\nu/c$. The number of states per volume, per unit frequency per solid angle is then,

$$\frac{1}{V}\frac{dN}{d\nu d\Omega} = m\frac{\nu^2}{c^3}.$$

Since there are independent polarizations, $m = 2$, and the total density of states is typically written as,

$$\frac{2\nu^2}{c^3}$$

to account for both polarizations.

4.6.5.3 Occupation Number

The final factor describing blackbody radiation is the average occupation number for a given state. Photons are bosons, and thus behave Bose–Einstein statistics, generally. However, in most situations, photons are free to be absorbed or emitted in their thermalized surroundings, which leads to a zero chemical potential, μ. Thus, the probability of finding a photon in an energy state s is

$$\mathrm{e}^{-\beta n_s h\nu}, \tag{4.26}$$

where $\beta = 1/k_B T$, T is the physical temperature, and k_B is Boltzmann's constant. The average occupation number can be computed as follows:

$$\bar{n}_s = \frac{\sum_{n_s=0}^{\infty} n_s \mathrm{e}^{-\beta n_s h\nu}}{\sum_{n_s=0}^{\infty} \mathrm{e}^{-\beta n_s h\nu}} = -\frac{1}{h\beta}\frac{\partial}{\partial\nu}\ln\left(\sum_{n_s=0}^{\infty}\mathrm{e}^{-\beta n_s h\nu}\right) = -\frac{1}{h\beta}\frac{\partial}{\partial\nu}\ln\left(1 - \mathrm{e}^{-\beta h\nu}\right)^{-1}, \tag{4.27}$$

$$\bar{n}_s = \frac{1}{\mathrm{e}^{h\nu/k_B T} - 1}. \tag{4.28}$$

4.6.5.4 The Planck Function

Putting these three factors together gives us the energy per volume per solid angle per frequency,

$$u_\nu(T) = (h\nu)\frac{2\nu^2}{c^3}\frac{1}{e^{h\nu/k_BT} - 1} = \frac{2h\nu^3}{c^3}\frac{1}{e^{h\nu/k_BT} - 1}. \tag{4.29}$$

If we imagine that we place a detector within the blackbody radiation and ask how much energy per time crosses the area of the detector within this solid angle in a given frequency range, then the volume we collect of the energy density is $dV = cdtdA$. Thus the specific intensity of the radiation, $B_\nu(T)$ is simply $cu_\nu(T)$, where c is the speed of light.

$$B_\nu(T) = \frac{2h\nu^3}{c^2}\frac{1}{e^{h\nu/k_BT} - 1}. \tag{4.30}$$

An illustration of the Planck function for selected values of T spanning a relevant astrophysical range is shown in Figure 4.19.

For most cases, noise limits are derived from foreground sources, since they are usually brighter than the small, polarized signal. Such photon sources are considered

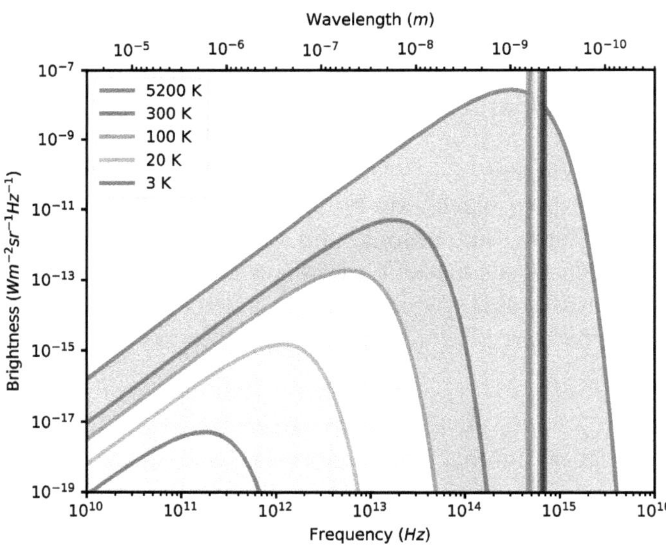

Figure 4.19. The Planck function is shown for a variety of temperatures. The approximate location of the visible spectrum is indicated by the rainbow stripe.

[3] Sometimes called *greybodies*.

to be modified blackbodies.[3] Thus, the emission spectrum of such sources is characterized by temperature, T, and the frequency-dependent emissivity, $\varepsilon(\nu)$.

$$I_\nu(T) = \varepsilon(\nu)B_\nu(T), \tag{4.31}$$

where $B_\nu(T)$ is the Planck function as derived above. Both $B_\nu(T)$ and $I_\nu(T)$ have units of specific intensity. In MKS units this corresponds to [$\mathrm{Wm^{-2}\ sr^{-1}\ Hz^{-1}}$].

4.6.5.5 Stefan–Boltzmann Law

We wish to find the total power emitted per area. To do this we assume the emitting surface is flat and that the emission occurs on one side of this surface. The surface is assumed to be Lambertian, so the intensity drops as $\cos\theta$, where θ is the polar angle measured from the vertical. If we integrate the intensity ($B_\nu(T)$) over all frequencies and all solid angles of the hemisphere, we find the power density per unit area of the emitter to be

$$|\mathrm{S}| = \sigma T^4.$$

Here, σ is the Stefan–Boltzamann constant and equal to

$$\sigma = \frac{2k^4\pi^5}{15h^3c^2} = 5.67 \times 10^8\ \mathrm{Wm^{-2}\ K^{-4}}.$$

4.6.6 Statistical Noise

Measurements are fundamentally limited by statistical fluctuations. These are due to quantum mechanical variations, and contributions to noise can occur in the photon fields of the source and/or foreground and from the detector and readout. The noise due to the detection system is dependent on the type of detectors used. Because of this and the fact that good polarimeter design dictates that the instrumental noise be subdominant to the photon noise in the system, here we will consider photon noise as the only contributor.[4] (Mather 1982) presents a review of the different types of noise sources commonly present in astronomical instruments.

To obtain an expression for the noise equivalent power for photon noise from a source, we start with the variance in the number of photons that a detector receives. Here, we will assume that the detector has an étendue, $A\Omega$, a small range of frequencies, $\delta\nu$, and an integration time t_{int}. This is given by Lamarre (1986) as

$$\sigma_{\bar{n}}^2 = \bar{n} + \frac{1}{g}\bar{n}^2. \tag{4.32}$$

This expression represents the fluctuation in the number of photons detected, given a mean detection of $\bar{n}$ from the source. The first term is due to so-called "photon shot noise." This is sometimes thought of as the "particle term" and dominates when $\bar{n} \ll 1$, which is typical at wavelengths shortwards of the spectrum's peak. The

[4] When the photon noise from the source is the dominant source of noise in a polarimeter/photometer, we say that the system is *background limited*, even though the photons may technically be coming predominantly from emission from sources that are between the source and the detector, or so-called "foregrounds."

second term is the "photon bunching" or "wave term." This term dominates at high occupation number ($\bar{n} \gg 1$). The factor $1/g$ that accompanies this term represents that for a multimode detector integrating for longer than a single coherence time, there will be multiple copies of the signal that cause the noise to be reduced by $\sqrt{1/g}$. In this case, we can write an expression for g, which is simply a counting of the number of independent bunches of photons that are detected,

$$g = m\frac{A\Omega}{\lambda^2}\frac{t_{\text{int}}}{\Delta\tau_c} = m\frac{A\Omega\nu^2}{c^2}t_{\text{int}}\delta\nu. \tag{4.33}$$

The factor m represents the number of polarizations present in the signal. The second factor is the number of spatial modes (1 mode has an étendue of $\sim\lambda^2$). The third factor is the length of the integration time in units of correlation time. Substituting, one obtains

$$\sigma_{\bar{n}}^2 = \bar{n} + \frac{c^2}{mA\Omega\nu^2 t_{\text{int}}(\delta\nu)}\bar{n}^2. \tag{4.34}$$

To find the contribution to the noise equivalent power from the photons within $\delta\nu$, we note that

$$\delta\text{NEP}^2 = \frac{(h\nu)^2\sigma_{\bar{n}}^2}{Bt_{\text{int}}^2} = \frac{2(h\nu)^2\sigma_{\bar{n}}^2}{t_{\text{int}}} = \frac{2(h\nu)^2}{t_{\text{int}}}\left[\bar{n} + \frac{c^2}{mA\Omega\nu^2(\delta\nu)}\bar{n}^2\right]. \tag{4.35}$$

Here, $B = 1/(2t_{\text{int}})$ is the measurement bandwidth. We can also relate the photon number to the specific intensity of the radiation,

$$\bar{n} = \alpha\frac{I_\nu(T)}{h\nu}(A\Omega)(\delta\nu)t_{\text{int}} = \alpha\varepsilon\frac{B_\nu(T)}{h\nu}(A\Omega)(\delta\nu)t_{\text{int}} \equiv \eta\frac{B_\nu(T)}{h\nu}(A\Omega)(\delta\nu)t_{\text{int}}. \tag{4.36}$$

Here, α is the quantum efficiency of the detector and optics. Because the quantum efficiency and the emissivity of the source enter the math in the same way, we define $\eta \equiv \alpha\varepsilon$ for convenience. Putting this all together, we find,

$$\delta\text{NEP}^2 = \frac{2(h\nu)^2}{t_{\text{int}}}\left[\left(\eta\frac{B_\nu(T)}{h\nu}(A\Omega)(\delta\nu)t_{\text{int}}\right) + \frac{c^2}{m(A\Omega)(\delta\nu)\nu^2 t_{\text{int}}}\left(\eta\frac{B_\nu(T)}{h\nu}(A\Omega)(\delta\nu)t_{\text{int}}\right)^2\right] \tag{4.37}$$

$$\delta\text{NEP}^2 = 2h\nu\eta B_\nu(T)(A\Omega)(\delta\nu)\left[1 + \frac{c^2\eta B_\nu(T)}{m(h\nu)\nu^2}\right]. \tag{4.38}$$

We substitute the expression above for $B_\nu(T)$,

$$\delta\text{NEP}^2 = 4(A\Omega)(h\nu)^2\eta\frac{\nu^2}{c^3}\frac{\delta\nu}{e^{h\nu/k_BT} - 1}\left(1 + \frac{2\eta/m}{e^{h\nu/k_BT} - 1}\right). \tag{4.39}$$

The total NEP for a finite frequency band, $\Delta\nu$ can be found by adding the squares of the δNEP from each of the (effectively) infinitesimal frequency intervals. If we take limit $\delta\nu \rightarrow d\nu$ and then integrate over all frequencies in the band, we recover the expression for NEP (Lamarre 1986),

$$NEP^2 = 4A\Omega \int_{\Delta\nu} (h\nu)^2 \eta \frac{\nu^2}{c^2} \frac{d\nu}{e^{h\nu/k_B T} - 1} \left(1 + \frac{2\eta/m}{e^{h\nu/k_B T} - 1}\right). \tag{4.40}$$

Here, η is the product of source emissivity (ε), cold transmission of the instrument, and detector quantum efficiency, while m is the number of polarizations being detected (1 or 2), and $A\Omega$ is the throughput of the optical system (étendue).

It is also sometimes convenient to work in unitless quantities in which one defines the scaled frequency $x \equiv \frac{h\nu}{k_B T}$ ($dx = \frac{h}{k_B T} d\nu$) (Richards 1994; Mather 1982). The integral can then be evaluated over the limits of interest and the scaling with physical parameters can be readily identified. In this case, Equation (4.40) becomes

$$\mathrm{NEP}^2 = 4\frac{A\Omega}{c^2} \frac{(k_\mathrm{B} T)^5}{h^3} \int_{\Delta x} \frac{x^4 dx}{e^x - 1} \left(1 + \frac{2\eta/m}{e^x - 1}\right)\eta. \tag{4.41}$$

Zmuidzinas (2003) and Lamarre (1986) point out that the second term in Equations (4.40) and (4.41) is an approximation that involves the assumption that all spatial modes are equally populated, and thus the uncertainties can be independently summed. If this is not the case, the correlations of Hanbury Brown and Twiss must be taken into account and this term increases (Brown & Twiss 1957, 1958a, 1958b, 1958c). Most of our use cases involve observations of extended sources where the intensity fills the telescope beam, Equation (4.40) is a good approximation. For observation of point sources, where the angular extent of the object is much smaller than the beam, such correlations need to be considered.

4.7 Calibration and Systematic Considerations

4.7.1 Instrument Stability

An additional complication is that the detectors, associated electronics, as well as other polarimeter components have limits to their stability. For example, in addition to stochastic noise sources such as phonon and Johnson noise, the detector responsivity (or the multiplicative factor that determines the ratio of the incident power to the reported voltage) is generally time-variable as relevant parameters in the system change. These sources of variability can be intrinsic to a given device structure or extrinsic (i.e., associated with environmental conditions which include vibration, ionizing radiation, electromagnetic fields, temperature, etc.). These effects can lead to correlated signatures, which, if unmitigated in the hardware or unaccounted for in calibration, can lead to excess noise and potential systematic artifacts in the instrument response and as a polarization bias in the subsequent data products.

4.7.2 Instrumental Polarization Correction

The motivation for measuring polarization is to probe the physical process by which the light became polarized. For this to occur, the source must exhibit some sort of anisotropy. Interstellar magnetic fields' effect on the alignment of dust grains provide the anisotropy for polarized emission from Galactic dust, and the geometry of Thomson scattering in the landscape of the inhomogeneity of the primordial density distribution provide the anisotropy for the polarization of the cosmic microwave background. In the same way, anisotropy in the light path of the instrument can also produce polarization. It is essential to separate any instrumental contribution to the polarization.

Instrumental polarization refers to the polarization imparted to the incoming light by the instrument. Off-axis reflection and refraction induce polarization because of the boundary conditions for the two polarization are different, leading to different Fresnel coefficients for two orthogonal polarizations: one that is perpendicular to the plane of incidence (also known as a *TE* wave because the electric field vector is parallel to the interface plane) and the other that is parallel to the plane of incidence (also known as a *TM* wave because the magnetic field vector is parallel to the plane of the interface).

Often materials are "biattenuent" meaning that the two polarizations have different absorption through the material. Even materials that are isotropic can manifest biattenuance for beam paths that are not parallel to the optical axis.

Instrumental polarization is usually dominated by coupling between Stokes I and the Q and U parameters. However, we can derive the general expression for a model of instrumental polarization using the Jones/Müller formalism. Let the attenuation of the electric field along two orthogonal directions in a linear basis be represented by α_1 and α_2, respectively. Also, assume that this basis is oriented at an angle θ with respect to the instrument reference frame. The Jones matrix for such a condition, which is

$$\mathsf{J} = \begin{pmatrix} \cos\theta & -\sin\theta \\ \sin\theta & \cos\theta \end{pmatrix} \begin{pmatrix} \alpha_1 & 0 \\ 0 & \alpha_2 \end{pmatrix} \begin{pmatrix} \cos\theta & \sin\theta \\ -\sin\theta & \cos\theta \end{pmatrix}, \tag{4.42}$$

$$\mathsf{J} = \begin{pmatrix} \alpha_1\cos^2\theta + \alpha_2\sin^2\theta & \alpha_1\sin\theta\cos\theta - \alpha_2\sin\theta\cos\theta \\ \alpha_2\cos\theta\sin\theta - \alpha_2\cos\theta\sin\theta & \alpha_1\sin^2\theta + \alpha_2\cos^2\theta \end{pmatrix}. \tag{4.43}$$

Here, we have assumed that the elements α_1 and α_2 are both real quantities. The generic case where these elements are complex can be readily derived. The corresponding Müller matrix for this situation is

$$\mathsf{M} = \begin{pmatrix} \frac{\alpha_1^2+\alpha_2^2}{2} & \frac{(\alpha_1^2-\alpha_2^2)}{2}\cos 2\theta & \frac{(\alpha_1^2-\alpha_2^2)}{2}\sin 2\theta & 0 \\ \frac{(\alpha_1^2-\alpha_2^2)}{2}\cos 2\theta & \frac{\alpha_1^2+\alpha_2^2}{2}\cos^2 2\theta + \alpha_1\alpha_2\sin^2 2\theta & \frac{(\alpha_1-\alpha_2)^2}{4}\sin 4\theta & 0 \\ \frac{(\alpha_1^2-\alpha_2^2)}{2}\sin 2\theta & \frac{(\alpha_1-\alpha_2)^2}{4}\sin 4\theta & \frac{\alpha_1^2+\alpha_2^2}{2}\sin^2 2\theta + \alpha_1\alpha_2\cos^2 2\theta & 0 \\ 0 & 0 & 0 & \alpha_1\alpha_2 \end{pmatrix}. \tag{4.44}$$

The measurement of one of the Stokes parameters, *e.g.*, Q, can then be written in terms of the incident Stokes parameters,

$$Q' = I\frac{(\alpha_1^2-\alpha_2^2)}{2}\cos 2\theta + Q\left\{\frac{\alpha_1^2+\alpha_2^2}{2}\cos^2 2\theta + \alpha_1\alpha_2\sin^2 2\theta\right\} + U\frac{(\alpha_1-\alpha_2)^2}{4}\sin 4\theta. \tag{4.45}$$

If we work in the limit where instrumental polarization is small, such that $\alpha_1,\ \alpha_2 \sim 1$ and that the observed polarization signature is small (this is reasonable since dust polarization is in almost all cases <20%.) In this case, the third term can be ignored since both U and $(\alpha_1 - \alpha_2)^2 \ll 1$. If we then divide the equation by I, we find,

$$q' = q_{\text{inst}} + q_{\text{sky}}, \tag{4.46}$$

where $q_{\text{inst}} = \frac{1}{2}(\alpha_1^2 - \alpha_2^2)\cos 2\theta$ is the instrumental contribution to the normalized Stokes parameters.

We can see here that for small instrumental polarization, the normalized Stokes parameters for the sky and instrumental effects are additive. This provides a convenient means for removal of this effect as utilized by Platt et al. (1991). For most modern polarimeters, this condition is met, as instrumental polarization contributions are typically only a few per cent. For example, the HAWC+ polarimeter on SOFIA has an instrumental polarization of ~2%, which is dominated by polarization-dependent absorption the off-axis reflection from the tertiary mirror (Harper et al. 2018).

Polarization artifacts can also originate from emission associated with instrument components and surfaces, which must also be taken into account in the calibrations. We refer the reader to Chapter 3.2.1 for a discussion of such an effect for reflective modulator.

4.7.3 Polarization Efficiency and Isolation

Polarization efficiency is the ratio of measured polarization to input polarization; *polarization isolation* is the level of mixing between two orthogonal polarization channels in a component or instrument system. For example, for the case of a wire grid that is designed to pass horizontal polarization (say, in the $\hat{\mathbf{y}}$-direction). The output power in the component of polarization parallel to this axis is

$$E_y^{2'} = \varepsilon E_y^2 + \eta E_x^2. \tag{4.47}$$

Here, ε is the polarization efficiency and η is the polarization isolation.

In many cases, especially those in which a polarization modulator is included in the optical path, the polarization efficiency is dominated by how well the modulator approximates the unitary (ideal) case. As an example we look at single-layer half-wave plates. For the design frequency or wavelength, the differential phase delay is exactly π; however, for a finite bandwidth, this value drifts as one moves either to larger or smaller wavelengths. The superposition of these frequencies is detected, resulting in a detected polarization that is smaller than the actual.

A simple method for measuring polarization efficiency is by observing a source of known polarization and then defining the efficiency to be the ratio of the measured to source polarization. The measured polarization of an unknown source then must be divided by the polarization efficiency, with an appropriate noise penalty. Alternately, the polarization modulation efficiency can be assessed by examining the noise penalty when including the modulation. The white noise level after demodulation will increase proportional to the inverse of the efficiency. Taking the ratio of the white noise level before and after demodulation can provide a measurement of the modulation efficiency (Harrington et al. 2021).

4.7.4 Co-polarization and Cross-polarization

Polarization efficiency and isolation are useful concepts for an instrument where it is safe to disregard angular dependence– for example in a transmission line or in a slow (high $f/$) optical system where the paraxial approximation holds. However, in general, the leakage between polarizations is often a function of angle and position within the beam. This drives the need to calibrate and understand the polarimeter angular response. In other words, for a given instrument element (sensor), what is the polarization response of as a function of angle on the sky? In fact, there are several factors at play here, including the current distribution in the sensor, how the far-field transform of those fields project through fore-optics (e.g., reflectors), and the projection onto a relevant angular basis on the sky. For cases where system requirements dictate high precision (and for fast optical systems), a full analysis of the instrument is typically required to fully parameterize and model the observed response.

Co-polarization and *cross-polarization* are the concepts of interest here and are typically expressed as a function of a polar coordinate system with the polar angle, θ, measured from the axis parallel to the beam. The azimuthal angle, ϕ, is measured from the x-axis, and the polarization direction in the x–y plane is assumed to be oriented in the y-direction. In contrast to efficiency and isolation, these quantities are expressed in terms of the electric fields as opposed to power.

The first step is to define a set of basis functions for these quantities. These functions will correspond to the directions of the co- and cross-polarization as a function of angular location on the sky and are functions of θ and ϕ The most commonly used definitions for co- and cross-polarization are introduced as the third definition in Ludwig (1973),

$$\begin{aligned}\hat{\mathbf{e}}_{\text{co}-\text{pol}} &\equiv \sin\phi\hat{\mathbf{e}}_\theta + \cos\phi\hat{\mathbf{e}}_\phi \\ &= -(1 - \cos\theta)\sin\phi\cos\phi\hat{\mathbf{x}} + [1 - \sin^2\phi(1 - \cos\theta)]\hat{\mathbf{y}} - \sin\theta\sin\phi\hat{\mathbf{z}}\end{aligned} \tag{4.48}$$

$$\begin{aligned}\hat{\mathbf{e}}_{\text{x}-\text{pol}} &\equiv \cos\phi\hat{\mathbf{e}}_\theta - \sin\phi\hat{\mathbf{e}}_\phi \\ &= [1 - \cos^2\phi(1 - \cos\theta)]\hat{\mathbf{x}} - (1 - \cos\theta)\sin\phi\cos\phi\hat{\mathbf{y}} - \sin\theta\cos\phi\hat{\mathbf{z}}.\end{aligned} \tag{4.49}$$

These unit vectors can be used to define the co- and cross-polarization components of the electric field vector as follows,

$$\begin{aligned}E_{\text{co}-\text{pol}} &= \mathbf{E}\cdot\hat{\mathbf{e}}_{\text{co}-\text{pol}} \\ E_{\text{x}-\text{pol}} &= \mathbf{E}\cdot\hat{\mathbf{e}}_{\text{x}-\text{pol}}.\end{aligned} \tag{4.50}$$

These components form two *beam patterns* for a telescope that are mutually orthogonal and provide a complete basis such that for any given direction, the polarization to which the detector is sensitive can be written as a linear combination of these two basis vectors. The bases for co- and cross-polarization are shown in Figure 4.20.

The term *E-plane* corresponds to the plane containing the beam (z) axis and is parallel to the co-polarization direction. In the geometry above, this corresponds to

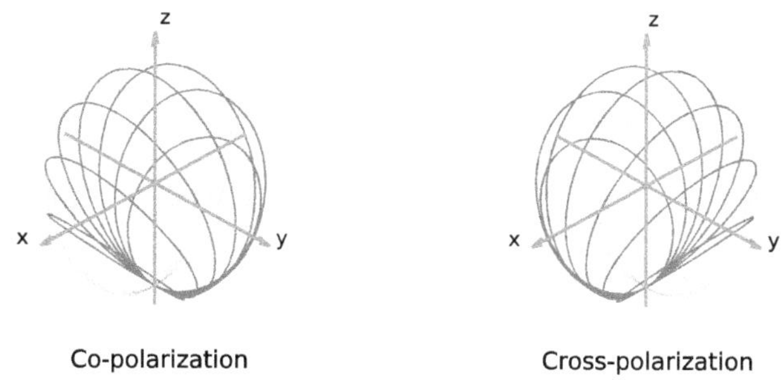

Figure 4.20. The (Ludwig III) bases for the co- (left; blue) and cross-polarization (right; red) are shown for a detector oriented for polarization in the y-direction. The blue and red curves sample the direction of the basis vectors of the co- and cross-polarization, respectively, as a function of the direction relative to the origin. In each case, the complementary beam pattern is shown in a lighter shade to highlight the orthogonality of the bases.

$x = 0$ or $\phi = \pi/2$. Correspondingly the *H-plane* corresponds to the orthogonal plane, $y = 0$ or $\phi = 0$. The difference of the beam profile between these two planes in intrinsically related to the cross-polarization response.

As an example, we will look at a calculation for a rotationally-symmetric antenna structure that is excited by a transverse current distribution on its symmetry axis. In this case, the dominant azimuthal variation in the resultant pattern will be sinusoidal. Representative examples include smooth wall TE_{11} feeds, corrugated HE_{11} feedhorns, hybrid Potter feeds, etc., and the beam response can be written as

$$\mathbf{E}(\theta, \phi) = E_E(\theta)\sin\phi\hat{\mathbf{e}}_\theta + E_H(\theta)\cos\phi\hat{\mathbf{e}}_\phi. \tag{4.51}$$

The solid angle of the beam can then be calculated, using the E and H plane beam profiles,

$$\Omega_{\text{beam}} = \iint_{4\pi} \left|\mathbf{E}(\theta, \phi)\right|^2 d\Omega \tag{4.52}$$

$$= \frac{1}{2}\int [|E_E(\theta)|^2 + |E_H(\theta)|^2] d(\cos\theta) \tag{4.53}$$

$$= \frac{1}{2}\int [|E_{\text{co-pol}}(\theta)|^2 + |E_{\text{x-pol}}(\theta)|^2] d(\cos\theta) \tag{4.54}$$

$$= \frac{4\pi}{G} = \frac{\lambda^2}{A_{\text{eff}}}, \tag{4.55}$$

where G is the antenna *gain* and A_{eff} is the effective area of the antenna. We notice that the étendue, $A\Omega = \lambda^2$, is as expected for a single mode antenna. From Equation (4.54) it can be seen that if the solid angle contribution from the cross-pol is significant relative to the main beam, it will need to be accounted for in the calibration or the flux scale will be incorrectly specified. Similarly, this response can couple to polarized foregrounds over a finite region, which will become of greater importance as systems push to higher sensitivity. For simplicity a rotationally symmetric beam has been adopted, however, similar considerations hold for more general antenna response profiles.

Given the E- and H-plane response, we find the co- and cross-polar amplitudes from Equation (4.50),

$$\begin{aligned} E_{\text{co-pol}} &= \mathbf{E}(\theta, \phi) \cdot \hat{\mathbf{e}}_{\text{co-pol}} \\ &= (E_E(\theta)\sin\phi\hat{\mathbf{e}}_\theta + E_H(\theta)\cos\phi\hat{\mathbf{e}}_\phi) \cdot (\sin\phi\hat{\mathbf{e}}_\theta + \cos\phi\hat{\mathbf{e}}_\phi) \\ &= E_E(\theta)\sin^2\phi + E_H(\theta)\cos^2\phi \\ &= \frac{1}{2}[E_E(\theta) + E_H(\theta)] - \frac{1}{2}[E_E(\theta) - E_H(\theta)]\cos 2\phi \end{aligned} \tag{4.56}$$

$$\begin{aligned} E_{\text{x-pol}} &= \mathbf{E}(\theta, \phi) \cdot \hat{\mathbf{e}}_{\text{x-pol}} \\ &= (E_E(\theta)\sin\phi\hat{\mathbf{e}}_\theta + E_H(\theta)\cos\phi\hat{\mathbf{e}}_\phi) \cdot (\cos\phi\hat{\mathbf{e}}_\theta - \sin\phi\hat{\mathbf{e}}_\phi) \\ &= \frac{1}{2}[E_E(\theta) - E_H(\theta)]\sin 2\phi. \end{aligned} \tag{4.57}$$

Figure 4.21 shows an example of co- and cross-polarization patterns from a HE_{11} feedhorn. In this case, the cross-polarization has been multiplied by 10 to aid in visualization.

Cross-polarization is a more significant consideration for optical elements that are an electrically-small number of wavelengths in size. In this limit, the field confinement is relatively poor, diffraction to large angles increases the importance of projection effects, and the differences between E-/H-boundaries require a vector treatment to capture the final polarization state. As a counterexample, if one assumes an electrically large system, significant beam power only occurs for $\theta \ll 1$, the equations in (4.58) reduce to

$$\hat{\mathbf{e}}_{\text{co-pol}} = \hat{\mathbf{y}} \tag{4.58}$$

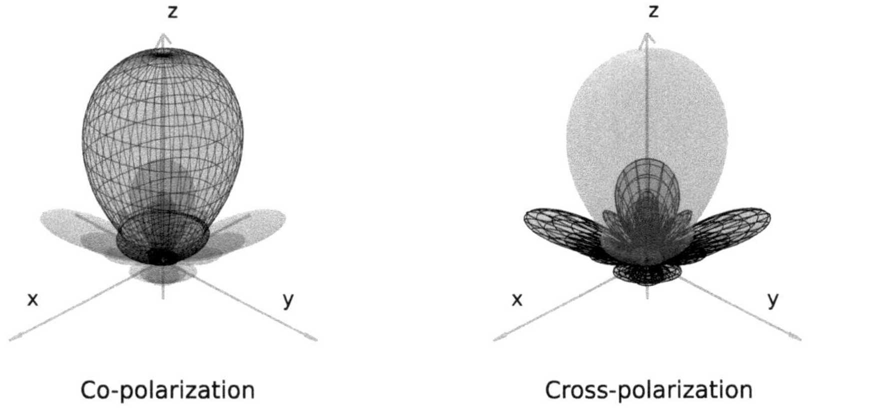

Figure 4.21. (Left) Co-polarization of an HE_{11} feedhorn is shown as a shaded blue wireframe pattern with the corresponding cross-polarization shown lightly in red. The cross-polarization is magnified by a factor of 10 for visibility. (Right) Same image as in the left side, but the cross-polarization is highlighted by the wireframe.

$$\hat{\mathbf{e}}_{\text{x-pol}} = \hat{\mathbf{x}}. \tag{4.59}$$

In this limit, the co- and cross-polarization coordinate system is approximately Cartesian, and there is little geometric mixing between the polarizations. This underscores the geometric origin of cross-polarization.

For polarimetry systems, beam shape and cross-polarization are both potential systematic concerns. Asymmetric beams can map unpolarized sub-beam structure into polarization in instruments that rely on the rotation of the instrument to cover the linear polarization space. This is especially worrisome for the case of bright compact sources. Modulators such as half-wave plates can mitigate this effect by decoupling the beam rotation from the polarization modulation. Cross-polarization can cause misidentification of sub-beam polarization into the signal of interest. In both cases, implementation of optical systems that have low cross-polarization and symmetric beams can help. The extent to which further mitigation may be necessary is dependent on the details of the instrument as well as its science goals.

4.7.5 Polarization Angle Calibration

One of the challenges of a polarimetry system is the calibration of the polarization angle. The precision to which this must be done depends on relevant science goals. For example, for the CMB, angle offsets can cause leakage from E modes into B modes. Given the large amplitude of EE relative to BB, even a small error in the polarization angle can greatly impact the power of the measurement.

Calibration involves mapping the linear polarization detection basis from the sky onto the detectors (and ultimately the processed maps). A theoretical understanding of how these coordinates are tracked through the system is crucial, especially since, depending on the system, instrument, telescope, and sky rotation all can be time-variable factors. While calculable in that the errors can be assessed and quantified (Murphy et al. 2024), the need for external verification become a more pressing concern as systems and measurements press to achieve higher sensitivities using a larger number of detectors.

Calibration can be achieved via observations. This can be through measurement of known astronomical sources; however, the precision of this technique is limited to previous knowledge (Kaufman et al. 2016). Keating et al. (2013) have devised a calibration technique that calibrates the polarization angle via the assumption that the EB and TB correlations should be zero. This is a precise method for self calibration of CMB polarization data, but it is model-dependent.

Specific testing to measure the polarization angle reference can be experimentally carried out using a polarized source. Often this is done by placing a sparse polarizing grid at an appropriate location in the optical path (Bradford et al. 2012; Murata et al. 2023), employing the polarized emission from a rotating mirror as a small signal source (Bischoff et al. 2013; Hasegawa et al. 2011), or via signal injection into a dual-mode waveguide structure (Peverini et al. 2006). In a similar spirit, a thin dielectric sheet can be used as the source of polarized emission (O'Dell et al. 2002). From a metrology perspective, ideally, this calibration source would be common to all the polarimetric sensors in the system. Recently, drone-based modulated calibration sources have been implemented (Dünner et al. 2020; Coppi et al. 2025) to provide precise calibration targets for ground-based instruments. Suborbital (Nati et al. 2017) and satellite (Johnson et al. 2015) implementation concepts have been reported in the literature.

References

Abitbol, M. H., Ahmed, Z., Barron, D., et al. 2017, arXiv:1706.02464 2017arXiv170602464A
Auton, J.P. 1967, ApOpt, 6, 1023
Barney, R. D., & Magner, T. J. 1988, Proc. SPIE, 973, 139
Bird, G. R., & Parrish Jr, M. 1960, JOSA, 50, 886
Bischoff, C., Brizius, A., Buder, I., et al. 2013, ApJ, 768, 9
Bock, J., Lange, A., Matsuhara, H., et al. 1995, ApOpt, 34, 2268
Bradford, K. J., Kovac, J. M., & Vieregg, A. G.Keck Array Team 2012, AAS Meeting, 219, 422.28
Brown, R. H., & Twiss, R. Q. 1957, RSPSA, 242, 300
Brown, R. H., & Twiss, R. Q. 1958a, RSPSA, 243, 291
Brown, R. H., & Twiss, R. Q. 1958b, RSPSA, 248, 199
Brown, R. H., & Twiss, R. Q. 1958c, RSPSA, 248, 222
Bryan, S. A., Ade, P. A. R., Amiri, M., et al. 2010, Proc. SPIE, 7741, 77412B
Burnside, W., & Chuang, C. 2003, ITAP, 30, 790
Cameron, R. M. 1976, S&T, 52, 327
Chambers, W. G., Parker, E. T. J., & Costley, E. A. E. 1986, Infrared and Millimeter Waves Vol. 16: Electromagnetic Waves in Matter, Part III, ed. K. J. Button (Amsterdam, Elsevier) 77
Chattopadhyay, G., & Zmuidzinas, J. 1998, ITAP, 46, 736
Chuss, D. T., Ade, P. A. R., Benford, D. J., et al. 2010, Proc. SPIE, 7741, 77411P
Chuss, D. T., Ali, A., Amiri, M., et al. 2016, JLTP, 184, 759
Chuss, D. T., Benford, D. J., Walker, C., et al. 2004, Proc. SPIE, 5492, 1450
Chuss, D. T., Eimer, J. R., Fixsen, D. J., et al. 2014, RScI, 85, 064501
Chuss, D. T., Wollack, E. J., Moseley, S. H., & Novak, G. 2006, ApOpt, 45, 5107
Chuss, D. T., Wollack, E. J., Moseley, S. H., & Novak, G. 2006, ApOpt, 45, 5107
Chuss, D. T., Wollack, E. J., Henry, R., et al. 2012a, ApOpt, 51, 197
Chuss, D. T., Wollack, E. J., Pisano, G., et al. 2012b, ApOpt, 51, 6824
Coleman, H., Brown, R., & Wright, B. 2003, ITAP, 23, 817
Coppi, G., Dachlythra, N., Nati, F., et al. 2025, ApJS, 279, 30
Costley, A. E., Hursey, K. H., Neill, G. F., & Ward, J. M. 1977, JOSAA, 67, 979
Coughlin, K. P. 2018, PhD thesis, Univ. Michigan
Crawford, t. A., Hogg, D. A., & Hunt, L. 1961, BSTJ, 40, 1095
Crooke, J. A., & Hagopian, J. G. 1998, Proc. SPIE, 343, 30
Crowe, E. J., Bennett, C. L., Chuss, D. T., et al. 2012, ITAS, 23, 2500505
Datta, R., Brewer, M. K., Couto, J. D., et al. 2024, ApJS, 273, 26
Day, P. K., LeDuc, H. G., Mazin, B. A., Vayonakis, A., & Zmuidzinas, J. 2003, Natur, 425, 817
Dragone, C. 1982, ITAP, 30, 331
Dünner, R., Fluxá, J., Best, S., & Carrero, F. 2020, Proc. SPIE, 11453, 114532P
EBEX Collaboration, Aboobaker, A. M., Ade, P., Araujo, D., et al. 2018, ApJS, 239, 7
Eichhorn, W. L., & Magner, T. J. 1986, OptEn, 25, 541

Eimer, J. R., Li, Y., Brewer, M. K., et al. 2024, ApJ, 963, 92
Essinger-Hileman, T., Kusaka, A., Appel, J. W., et al. 2016, RScI, 87, 094503
Filippini, J. P., Ade, P. A. R., Amiri, M., et al. 2010, Proc. SPIE, 7741, 77411N
Fissel, L. M., Ade, P. A. R., Angilè, F. E., et al. 2010, Proc. SPIE, 7741, 77410E
Friberg, P., Bastien, P., Berry, D., et al. 2016, Proc. SPIE, 9914, 991403
Gandilo, N. N., Ade, P. A. R., Benford, D., et al. 2016, Proc. SPIE, 9914, 99141J
Gervasi, M., de Bernardis, P., De Petris, M., Masi, S., & Pisano, G. 1998, ApOpt, 37, 4277
Glenn, J., Walker, C. K., & Young, E. T. 1997, IJIMW, 18, 285
Goldsmith, P. 1998, Quasioptical Systems: Gaussian Beam Quasioptical Propagation and Applications (Piscataway, NJ: IEEE)
Gullett, I., Benson, B., Besuner, R., et al. 2023, ApOpt, 62, 4334
Gundersen, J., & Wollack, E. 2009, JPhCS, 155, 012005
Haghshenas, J., & Johari, E. 2014, Proc. SPIE, 9264, 92641R
Haller, E. E., Itoh, K. M., Beeman, J. W., Hansen, W. L., & Ozhogin, V. I. 1994, Proc. SPIE, 2198, 630
Harper, D. A., Runyan, M. C., Dowell, C. D., et al. 2018, JAI, 7, 1840008
Harrington, K., Datta, R., Osumi, K., et al. 2021, ApJ, 922, 212
Hasegawa, M., Tajima, O., Chinone, Y., et al. 2011, RScI, 82, 054501
Hinderks, J. R., Ade, P., Bock, J., et al. 2009, ApJ, 692, 1221
Hughes, D. H., Jáuregui Correa, J.-C., Schloerb, F. P., et al. 2010, Proc. SPIE, 7733, 773312
Hughes, D. H., Schloerb, F. P., Aretxaga, I., et al. 2020, Proc. SPIE, 11445, 1144522
Irwin, K. D. 2002, PhyC, 368, 203
Irwin, K. D., & Hilton, G. C. 2005, Cryogenic Particle Detection, Vol. 99 ed. C. Enss (New York: Springer) 63
Johnson, B. R., Vourch, C. J., Drysdale, T. D., et al. 2015, JAI, 4, 1550007
Kaufman, J. P., Keating, B. G., & Johnson, B. R. 2016, MNRAS, 455, 1981
Keating, B. G., Ade, P. A. R., Bock, J. J., et al. 2003, Proc. SPIE, 4843, 284
Keating, B. G., Shimon, M., & Yadav, A. P. S. 2013, ApJL, 762, L23
Komatsu, K., Ishino, H., Katayama, N., Matsumura, T., & Sakurai, Y. 2021, JATIS, 7, 034005
Koopman, B., Austermann, J., Cho, H.-M., et al. 2016, Proc. SPIE, 9914, 99142T
Krejny, M., Chuss, D., D'Aubigny, C. D., et al. 2008, ApOpt, 47, 4429
Kuo, C. L., Bock, J. J., Bonetti, J. A., et al. 2008, Proc. SPIE, 7020, 702011
Kusaka, A., Essinger-Hileman, T., Appel, J. W., et al. 2014a, RScI, 85, 039901
Kusaka, A., Wollack, E. J., & Stevenson, T. R. 2014b, JOSAA, 31, 1557
Lahtinen, J., & Hallikainen, M. 1999, Fabrication and Characterization of Large Free-standing Polarizer Grids for Millimeter Waves Vol. 20 (Berlin: Springer)
Lamarre, J. M. 1986, ApOpt, 25, 870
Landecker, T., Smegal, R., & McKinley, J. 1997, RaSc, 32, 2139
Lanting, T. M., Cho, H.-M., Clarke, J., et al. 2005, ApPhL, 86, 112511
Lawrence, C. R., Herbig, T., & Readhead, A. C. 1994, Proc. IEEE, 82, 763
Lazareff, B., Mahieu, S., & Geoffroy, D. 2009, Twentieth Int. Symp. on Space Terahertz Technology ed. E. Bryerton (Charlottesville, VA: University of Virginia) 223
Lee, D., Novak, G., Berthoud, M., et al. 2022, Proc. SPIE, 11453, 121901O
Leinert, C., & Klüppelberg, D. 1974, ApOpt, 13, 556
Leonardi, R., Williams, B., Bersanelli, M., et al. 2006, NewAR, 50, 977
Low, F. J. 1961, JOSA, 51, 1300

Ludwig, A. 1973, ITAP, 21, 116
Marcuvitz, N. 1951, Waveguide Handbook, MIT Rad. Labs. Ser. Vol. 10 (New York: McGraw-Hill)
Martin, D. H., & Puplett, E. 1970, InfPh, 10, 105
Mather, J. 2003, ITAP, 29, 967
Mather, J. C. 1982, ApOpt, 21, 1125
McMahon, J., Appel, J., Austermann, J., et al. 2009, AIP Conf. Proc., 1185, 490 (New York: AIP)
Mehta, C. L., & Wolf, E. 1964a, PhRv, 134, 1143
Mehta, C. L., & Wolf, E. 1964b, PhRv, 134, 1149
Mehta, C. L., & Wolf, E. 1967a, PhRv, 161, 1328
Mehta, C. L., & Wolf, E. 1967b, PhRv, 157, 1183
Mehta, C. L., Wolf, E., & Balachandran, A. P. 1966, JMP, 7, 133
Moyerman, S., Bierman, E., Ade, P. A. R., et al. 2013, ApJ, 765, 64
Murata, M., Nakata, H., Iijima, K., et al. 2023, RScI, 94, 124502
Murphy, C. C., Choi, S. K., Datta, R., et al. 2024, ApOpt, 63, 5079
Nati, F., Devlin, M. J., Gerbino, M., et al. 2017, JAI, 6, 1740008
Niemack, M. D. 2016, Appl. Opt., 55, 1688
Novak, G., Gonatas, D. P., Hildebrand, R. H., & Platt, S. R. 1989a, PASP, 101, 215
Novak, G., Sundwall, J. L., & Pernic, R. J. 1989b, ApOpt, 28, 3425
O'Brient, R., Ade, P., Arnold, K., et al. 2010, Proc. SPIE, 7741, 77410J
O'Dell, C. W., Swetz, D. S., & Timbie, P. T. 2002, ITMTT, 50, 2135
Padilla, I. L., Eimer, J. R., Li, Y., et al. 2020, ApJ, 889, 105
Pancharatnam, S. 1955, PIAS, 41, 137
Persky, M. 1999, RScI, 70, 2193
Petroff, M. A., Eimer, J. R., Harrington, K., et al. 2020, ApJ, 889, 120
Peverini, O. A., Tascone, R., Carretti, E., et al. 2006, ITMTT, 54, 412
Pisano, G. 2003, Proc. SPIE, 4843, 348
Pisano, G., Maffei, B., Ade, P. A. R., et al. 2016, ApOpt, 55, 10255
Pisano, G., Maffei, B., Ng, M. W., et al. 2014, Proc. SPIE, 9153, 915317
Pisano, G., Nesti, R., Ng, M., et al. 2012, JEWA, 26, 707
Pisano, G., Ritacco, A., Monfardini, A., et al. 2022, A&A, 658, A24
Pisano, G., Savini, G., Ade, P. A. R., Haynes, V., & Gear, W. K. 2006, ApOpt, 45, 6982
Platt, S. R., Hildebrand, R. H., Pernic, R. J., Davidson, J. A., & Novak, G. 1991, PASP, 103, 1193
Pompea, S. M. 1995, SSRv, 74, 181
Reif, F. 1965, Fundamentals of Statistical and Thermal Physics (New York: McGraw-Hill)
Richards, P. L. 1994, JAP, 76, 1
Rock, D., Warren, A., & Lewanski, A. 1982, Proc. SPIE, 330, 60
Rohrbach, S. O. 2023, Proc. SPIE, 12685, 24
Ruhl, J., Ade, P. A. R., Carlstrom, J. E., et al. 2004, Proc. SPIE, 5498, 11
Rybicki, G. B., & Lightman, A. P. 1979, Radiative Processes in Astrophysics (New York: Wiley)
Savini, G., Pisano, G., & Ade, P. A. R. 2006, ApOpt, 45, 8907
Shannon, C. E. 1949, IEEEP, 37, 10
Sharp, E., Benford, D., Fixsen, D., et al. 2012, Proc. SPIE, 8452, 1048
Shi, R., Brewer, M. K., Chan, C. Y. Y., et al. 2024, Proc. SPIE, 13102, 131021T
Siringo, G., Kreysa, E., Reichertz, L. A., & Menten, K. M. 2004, A&A, 422, 751

Sternberg, S. 1994, Group Theory and Physics (1st ed.; Cambridge: Cambridge Univ. Press)
Uher, J., Bornemann, J., & Rosenberg, U. 1993, Waveguide Components for Antenna Feed Systems: Theory and CAD (Norwood, MA: Artech House)
Voellmer, G. M., Bennett, C., Chuss, D. T., et al. 2008, Proc. SPIE, 7014, 70142A
Wollack, E. 2009, JPCS, 155, 42
Wollack, E. J., Kongpop, U., Chuss, D. T., et al. 2010, 2010 IEEE MTT-S Int. Microwave Symp. (New York: IEEE) 177
Zeng, L., Bennett, C. L., Chuss, D. T., & Wollack, E. J. 2010, ITAP, 58, 1383
Zmuidzinas, J. 2003, ApOpt, 42, 4989
Zmuidzinas, J. 2012, ARCMP, 3, 169
Zmuidzinas, J., & Richards, P. L. 2004, IEEEP, 92, 1597

Chapter 5

Polarimeters

Now that we have examined the elements of polarimetry, including instrumentation and related concepts, we turn our attention to polarimeter systems. In designing a polarimeter, there are many factors to consider. For example, depending on the observational frequency, technological approaches differ. At lower frequencies, the use of phase-sensitive detectors (e.g., mixers, correlators and amplifier based systems) have been employed. At higher frequencies, direct detectors such as NTD bolometers, TES bolometers, and MKIDs have found use for their sensitivity and scalability. In addition, space-borne or balloon-borne cameras have more constraints (in terms of, for example, mass, power, and volume) than do ground-based counterparts, but also face lower backgrounds, sources of ionizing radiation, and can achieve higher stability. This can motivate simpler signal chains for space-borne polarimeter implementations.

We begin this chapter with an overview of different polarimeter architectures. We follow this with a brief overview of the history of far-infrared and submillimeter polarimeters.

5.1 Polarimeter Architectures

Here, we explore a set of generic architectures in three distinct categories: (1) Coherent multiplier detection schemes, which involve a correlator on the back-end; (2) Switched phase-sensitive detection schemes, which utilize discrete phase switches in front of diplexers and simultaneous detection of two orthogonal polarizations; the detector elements in these cases can be coherent or total-power detectors; and (3) Continuous-parameter modulation schemes. We describe these systems without necessarily specifying a specific implementation. For example, a discrete phase switch could be implemented in the free-space optical path of the system, in a waveguide or in a microstrip circuit that is integrated with the detector. However, we give examples of the architectures where possible.

doi:10.1088/2514-3433/ae1022ch5

For this analysis we utilize a Jones matrix approach. Throughout this discussion, it is useful to recall that ideal polarization modulation is represented by unitary Jones matrices that correspond to rotations in Poincaré space. The representations of a collection of polarization circuit elements and their Jones matrix representation are given in Table 5.1. A summary of these architectures, their Jones topologies, and corresponding modulated measurements are given in Table 5.2. We provide two examples of the calculations to connect the architectures described by the Jones matrices to the modulation functions as examples. Throughout, we assume that the polarization basis (at the detector) is defined such that a difference in power in these two polarization defines the Stokes Q parameter. A correlation between the fields in these orientations then represents a measurement of Stokes U. The architecture

Table 5.1. Selected Jones Matrices

Element	Symbol	Jones Matrix
Phase Delay	$\bar{J}_{\Delta\phi}$	$\begin{pmatrix} 1 & 0 \\ 0 & \mathrm{e}^{\mathrm{i}\Delta\phi} \end{pmatrix}$
Gain or Attenuation	$\bar{J}_G$	$\begin{pmatrix} G_1 & 0 \\ 0 & G_2 \end{pmatrix}$
Orthomode Transducer (Linear Polarization Diplexer)	$\bar{J}_{\mathrm{OMT}}$	$\begin{pmatrix} 1 & 0 \\ 0 & 1 \end{pmatrix}$
0° In-Phase Power Divider	$\bar{J}_{\mathrm{OH}}$	$\frac{1}{\sqrt{2}}\begin{pmatrix} 1 & 1 \\ 1 & 1 \end{pmatrix}$
90° 3 dB Hybrid (Short-Slot Coupler)	$\bar{J}_{90H}$	$\frac{1}{\sqrt{2}}\begin{pmatrix} 1 & \mathrm{i} \\ \mathrm{i} & 1 \end{pmatrix}$
180° 3 dB Hybrid (Magic-Tee)	$\bar{J}_{180H}$	$\frac{1}{\sqrt{2}}\begin{pmatrix} 1 & 1 \\ 1 & -1 \end{pmatrix}$
Circular Polarizer Hybrid (Septum Polarizer)	$\bar{J}_{\mathrm{CPH}}$	$\frac{1}{\sqrt{2}}\begin{pmatrix} 1 & \mathrm{i} \\ \mathrm{i} & 1 \end{pmatrix}$
Ideal Multiplier/Correlator	$\bar{J}_{\times} = \bar{U}$	$\begin{pmatrix} 0 & 1 \\ 1 & 0 \end{pmatrix}$
Linear Polarized Detector (Horizontal)	$\bar{J}_{+}$	$\begin{pmatrix} D_+ & 0 \\ 0 & 0 \end{pmatrix}$
Linear Polarized Detector (Vertical)	$\bar{J}_{-}$	$\begin{pmatrix} 0 & 0 \\ 0 & D_- \end{pmatrix}$
Half-wave Plate	$\bar{J}_{\mathrm{HWP}}(\theta)$	$\begin{pmatrix} \mathrm{i}\cos(2\theta) & \mathrm{i}\sin 2\theta \\ \mathrm{i}\sin 2\theta & -\mathrm{i}\cos 2\theta \end{pmatrix}$
Quarter-Wave Plate	$\bar{J}_{\mathrm{QWP}}(\theta)$	$\frac{1}{\sqrt{2}}\begin{pmatrix} 1 + \mathrm{i}\cos 2\theta & i\sin 2\theta \\ \mathrm{i}\sin 2\theta & 1 - \mathrm{i}\cos 2\theta \end{pmatrix}$
Faraday Rotator	$\bar{J}_{\mathrm{FR}}(\Delta)$	$\begin{pmatrix} \cos\Delta & -\sin\Delta \\ \sin\Delta & \cos\Delta \end{pmatrix}$
Variable-Delay Polarization Modulator	$\bar{J}_{\mathrm{VPM}}(\theta, \Delta = 2\xi)$	$\begin{pmatrix} \cos\xi - \mathrm{i}\sin\xi\cos 2\theta & -\mathrm{i}\sin\xi\sin 2\theta \\ \mathrm{i}\sin\xi\sin 2\theta & -\cos\xi - \mathrm{i}\sin\xi\cos 2\theta \end{pmatrix}$
Rotating Wire Grid in Transmission	$\bar{J}_{\mathrm{WT}}(\theta)$	$\begin{pmatrix} \sin^2\theta & -\sin\theta\cos\theta \\ \sin\theta\cos\theta & -\sin^2\theta \end{pmatrix}$

Table 5.2. A Selection of Polarization Architectures

Example	Measurement Approach	Modulation Scheme	Lock-In Measurement
1	$\bar{J}_{\Delta\phi}\bar{J}_{G}\bar{J}_{OMT}\vert E_{in}\rangle$	$G_1G_2(U\cos\Delta\phi - V\sin\Delta\phi)$	$2G_1G_2U$
2	$\bar{J}_{\Delta\phi}\bar{J}_{G}\bar{J}_{90H}\bar{J}_{OMT}\vert E_{in}\rangle$	$G_1G_2(U\cos\Delta\phi - Q\sin\Delta\phi)$	$2G_1G_2U$
3	$\bar{J}_{\Delta\phi}\bar{J}_{G}\bar{J}_{180H}\bar{J}_{OMT}\vert E_{in}\rangle$	$G_1G_2(Q\cos\Delta\phi + V\sin\Delta\phi)$	$2G_1G_2Q$
4	$\bar{J}_{\Delta\phi}\bar{J}_{G}\bar{J}_{OMT}\bar{J}_{QWP}(0)\vert E_{in}\rangle$	$G_1G_2(V\cos\Delta\phi + U\sin\Delta\phi)$	$2G_1G_2V$
5	$\bar{J}_{180H}\bar{J}_{\Delta\phi}\bar{J}_{G}\bar{J}_{OMT}\vert E_{in}\rangle$	$\frac{D_\pm}{4}[I(G_1^2+G_2^2)+Q(G_1^2+G_2^2)\pm 2G_1G_2(U\cos\Delta\phi - V\sin\Delta\phi)]$	$G_1G_2(D_+ + D_-)U$
6	$\bar{J}_{180H}\bar{J}_{\Delta\phi}\bar{J}_{G}\bar{J}_{90H}\bar{J}_{OMT}\vert E_{in}\rangle$	$\frac{D_\pm}{4}[I(G_1^2+G_2^2)-V(G_1^2-G_2^2)\pm 2G_1G_2(U\cos\Delta\phi - Q\sin\Delta\phi)]$	$G_1G_2(D_+ + D_-)U$
7	$\bar{J}_{180H}\bar{J}_{\Delta\phi}\bar{J}_{G}\bar{J}_{180H}\bar{J}_{OMT}\vert E_{in}\rangle$	$\frac{D_\pm}{4}[I(G_1^2+G_2^2)+U(G_1^2-G_2^2)\pm 2G_1G_2(Q\cos\Delta\phi + V\sin\Delta\phi)]$	$G_1G_2(D_+ + D_-)Q$
8	$\bar{J}_{180H}\bar{J}_{\Delta\phi}\bar{J}_{G}\bar{J}_{OMT}\bar{J}_{QWP}(0)\vert E_{in}\rangle$	$\frac{D_\pm}{4}[I(G_1^2+G_2^2)+Q(G_1^2-G_2^2)\pm 2G_1G_2(V\cos\Delta\phi + U\sin\Delta\phi)]$	$G_1G_2(D_+ + D_-)V$
9	$\bar{J}_{90H}\bar{J}_{\Delta\phi}\bar{J}_{G}\bar{J}_{OMT}\vert E_{in}\rangle$	$\frac{D_\pm}{4}[I(G_1^2+G_2^2)+Q(G_1^2-G_2^2)\mp 2G_1G_2(V\cos\Delta\phi + U\sin\Delta\phi)]$	$-G_1G_2(D_+ + D_-)V$
10	$\bar{J}_{90H}\bar{J}_{\Delta\phi}\bar{J}_{G}\bar{J}_{90H}\bar{J}_{OMT}\vert E_{in}\rangle$	$\frac{D_\pm}{4}[I(G_1^2+G_2^2)-V(G_1^2-G_2^2)\mp 2G_1G_2(Q\cos\Delta\phi + U\sin\Delta\phi)]$	$-G_1G_2(D_+ + D_-)Q$
11	$\bar{J}_{90H}\bar{J}_{\Delta\phi}\bar{J}_{G}\bar{J}_{180H}\bar{J}_{OMT}\vert E_{in}\rangle$	$\frac{D_\pm}{4}[I(G_1^2+G_2^2)+U(G_1^2-G_2^2)\pm G_1G_2(V\cos\Delta\phi - Q\sin\Delta\phi)]$	$G_1G_2(D_+ + D_-)V$
12	$\bar{J}_{90H}\bar{J}_{\Delta\phi}\bar{J}_{G}\bar{J}_{OMT}\bar{J}_{QWP}(0)\vert E_{in}\rangle$	$\frac{D_\pm}{4}[I(G_1^2+G_2^2)+Q(G_1^2-G_2^2)\pm G_1G_2(U\cos\Delta\phi - V\sin\Delta\phi)]$	$G_1G_2(D_+ + D_-)U$
13	$\bar{J}_{0H}\bar{J}_{\Delta\phi}\bar{J}_{G}\bar{J}_{OMT}\vert E_{in}\rangle$	$\frac{D_\pm}{4}[I(G_1^2+G_2^2)+Q(G_1^2+G_2^2)+2G_1G_2(U\cos\Delta\phi - V\sin\Delta\phi)]$	$G_1G_2D_\pm U$
14	$\bar{J}_{0H}\bar{J}_{\Delta\phi}\bar{J}_{G}\bar{J}_{90H}\bar{J}_{OMT}\vert E_{in}\rangle$	$\frac{D_\pm}{4}[I(G_1^2+G_2^2)-V(G_1^2-G_2^2)+2G_1G_2(U\cos\Delta\phi - Q\sin\Delta\phi)]$	$G_1G_2D_\pm U$
15	$\bar{J}_{0H}\bar{J}_{\Delta\phi}\bar{J}_{G}\bar{J}_{180H}\bar{J}_{OMT}\vert E_{in}\rangle$	$\frac{D_\pm}{4}[I(G_1^2+G_2^2)+U(G_1^2-G_2^2)+2G_1G_2(Q\cos\Delta\phi + V\sin\Delta\phi)]$	$G_1G_2D_\pm Q$
16	$\bar{J}_{0H}\bar{J}_{\Delta\phi}\bar{J}_{G}\bar{J}_{OMT}\bar{J}_{QWP}(0)\vert E_{in}\rangle$	$\frac{D_\pm}{4}[I(G_1^2+G_2^2)+Q(G_1^2-G_2^2)+2G_1G_2(V\cos\Delta\phi + U\sin\Delta\phi)]$	$G_1G_2D_\pm V$
17	$\bar{J}_{FR}(\Delta)\vert E_{in}\rangle$	$\frac{D_\pm}{2}[I\pm(Q\cos 2\Delta - U\sin 2\Delta)]$	
18	$\bar{J}_{VPM}(\pi/4,\ \Delta)\vert E_{in}\rangle$	$\frac{D_\pm}{2}[I\pm(Q\cos 2\Delta + V\sin 2\Delta)]$	
19	$\bar{J}_{HWP}(\theta)\vert E_{in}\rangle$	$\frac{D_\pm}{2}[I\pm(Q\cos 4\theta + U\sin 4\theta)]$	
20	$\bar{J}_{QWP}(\theta)\vert E_{in}\rangle$	$\frac{D_\pm}{4}[2I\pm(Q+Q\cos 4\theta + U\sin 4\theta)\mp V\sin 2\theta]$	
21	$\bar{J}_{WT}(\theta)\vert E_{in}\rangle$	$\frac{D_\pm}{4}[I-(Q\cos 2\theta + U\sin 2\theta)]$	

Examples 1–4 are coherent multiplier detection schemes, 5–16 represent switched phase-sensitive detection schemes, and 17–21 are continuous-parameter modulation schemes.

between the detectors and the sky will alter the mapping of Stokes parameters from the sky into the detector basis.

5.1.1 Coherent Multiplier Detection Schemes

Entries 1-4 in Table 5.2 list four variations on correlator-based polarimeters. In each of these architectures, the detector measures the correlation between two orthogonal linear polarizations. This is equivalent to measuring the Stokes U parameter. Because of this, the detection element is represented by the Jones matrix having Stokes U as its eigenvalue. As will be illustrated in the following sections, this is in contrast to the case of incoherent detectors that have Stokes Q-like behavior, since in these systems, the two linear orthogonal polarizations are detected. These measurements can be calibrated and differenced, if desired.

For each case in this category, dual linear polarizations are accepted by the instrument and a controllable phase delay ($\Delta\phi$) is inserted into one of the signal paths. A gain (G_1, G_2) is inserted to enable representation of the asymmetry between the optical paths of the two polarizations. Various elements are inserted beyond the phase delay to change the polarization detection properties of the polarimeter from one example to the next. These devices include 180° hybrid couplers (180H), 90° hybrid couplers (90H), orthomode transducers (OMT), and a quarter-wave plate oriented at 0° with respect to the OMT axis.

As mentioned above, although we have given these devices labels that imply an exact implementation for each, the matrices can represent more general devices and could be conceivably implemented in a variety of ways either quasioptically, in microstrip, or in waveguide. For example, in Example 2, the architecture could have been equivalently written as $\bar{J}_{\Delta\phi}\bar{J}_{\mathrm{G}}\bar{J}_{\mathrm{CPH}}|E_{\mathrm{in}}\rangle$ or $\bar{J}_{\Delta\phi}\bar{J}_{\mathrm{G}}\bar{J}_{\mathrm{OMT}}\bar{J}_{\mathrm{QWP}}(\pi/4)|E_{\mathrm{in}}\rangle$ where $\bar{J}_{\mathrm{CPH}}$ represents the Jones matrix for a septum polarizer and $\bar{J}_{\mathrm{QWP}}(\pi/4)$ represents that of a quarter-wave plate oriented at 45° with respect to the OMT. The resulting product matrix is the same in each case.

Modulation in each of these cases is done by switching the value of $\Delta\phi$ from 0 to π and subtracting the two signals. The result of this lock-in measurement is shown in the final column of Table 5.2. It should be noted that Table 5.2 is complete in the sense that it explores architectures for all combinations of Q, U, and V to be modulated. Other architectures are possible, but they would only differ from these by a phase in the modulated function. This can be seen in the similarity between Examples 1 and 4.

5.1.2 Switched Phase-Sensitive Detection Schemes

Table 5.2 also shows a set of architectures based on phase switches (entries 5–16). In each case, both horizontal and vertical linear polarization are detected at the output of the system. The expressions for the signal in each indicate a detector gain of D_+ for the detector sensitive to horizontal polarization and D_- for the detector sensitive to vertical polarization. Note that the detector gains ($D_\pm$) are gains in power as opposed to $G_\pm$, which applied to fields. The final column is an expression for the difference in the signals between phase values of 0 and π. In examples 5–12, the horizontal and vertical detectors have been differenced after the lock-in demodulation. For Examples 13–16, the individual detector signals are given, as the output signals are copies of one another in these architectures.

Examples in this category (entries 5–16) in Table 5.2 are grouped according to the hybrid coupler element closest to the detectors. In the case of the first and third group, the polarization architectures are similar in that the 180° and 0° hybrid couplers maintain the linear polarization sensitivity of the detectors. The major difference between these two cases is that the 180° coupler inserts a phase flip in the vertical polarization detection channel that causes the two detectors to flip the sign of their modulated components relative to their baselines. This provides a natural path to baseline (or offset) modulation and stabilization in the system.

In the case of the models based on 90° hybrid couplers (Examples 9–12), the detectors are effectively circularly polarized. The different pattern of output Stokes parameters in the series of lock-in measurements is a direct result of this choice.

We repeat the point mentioned above, that because of the equivalence of their Jones matrices, an OMT followed by a 90° hybrid can be replaced either by a septum polarizer or a quarter-wave plate (QWP) at fixed at an angle of $\pi/4$, followed by an OMT, without changing the architecture. This relationship is quantified by

$$\bar{J}_{90H} \cdot \bar{J}_{\mathrm{OMT}} = \bar{J}_{\mathrm{CPH}} = \bar{J}_{\mathrm{OMT}} \cdot \bar{J}_{\mathrm{QWP}}(\pi/4). \tag{5.1}$$

We note that the architecture illustrated by Example 7 is that implemented by WMAP (Jarosik et al. 2003a, 2003b). The architecture for the Q/U Imaging ExperienT (QUIET) is similar to the devices shown here, but with additional power division to enable Q, U, and I measurement (Bischoff et al. 2013).

Example: Coherent Detection

To guide the reader through the architectures, we perform the calculation in Example 1 (Table 5.2). The layout is shown schematically in Figure 5.1 and consists of an OMT to separate the incoming signal into linear orthogonal polarizations, asymmetric gain terms to represent differential loss in the lines. We insert a variable phase $\Delta\phi$ in one of the lines that can be switched between values of 0 and π. Finally, the signals are combined in a correlator. We start the calculation by acting on the incoming signal that is represented by two orthogonal components, which we package into a column vector and represent generally by a "ket," $|E_{\mathrm{in}}\rangle$. The operators that represent the components of the polarimeter are applied to the incoming signal in the order they appear,

$$|E_{\mathrm{out}}\rangle = \bar{J}_{\Delta\phi}\bar{J}_{\mathrm{G}}\bar{J}_{\mathrm{OMT}}|E_{\mathrm{in}}\rangle. \tag{5.2}$$

We then represent each by their appropriate matrix from Table 5.1,

$$\begin{pmatrix} 1 & 0 \\ 0 & e^{i\Delta\phi} \end{pmatrix}\begin{pmatrix} G_1 & 0 \\ 0 & G_2 \end{pmatrix}\begin{pmatrix} 1 & 0 \\ 0 & 1 \end{pmatrix}\begin{pmatrix} E_x \\ E_y \end{pmatrix} = \begin{pmatrix} G_1 & 0 \\ 0 & G_2 e^{i\Delta\phi} \end{pmatrix}\begin{pmatrix} E_x \\ E_y \end{pmatrix}.$$

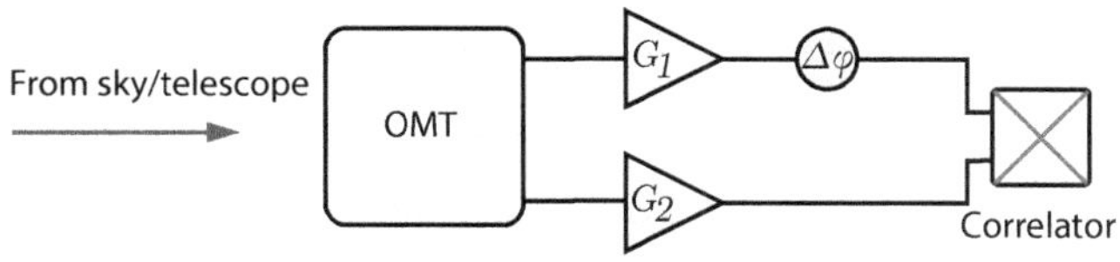

Figure 5.1. Schematic layout for Example 1 in Table 5.2, an architecture involving coherent detection using a correlator.

The detected (power) signal can be found by multiplying this signal by its complex conjugate with the operator for the detector element (in this case, the correlator, which is represented by the Stokes U matrix):[1]

$$\mathscr{P}_{\mathrm{detected}} = \langle E_{\mathrm{in}} | \bar{J}^{\dagger}_{\mathrm{OMT}} \bar{J}^{\dagger}_{G} \bar{J}^{\dagger}_{\Delta\phi} \bar{U} \bar{J}_{\Delta\phi} \bar{J}_{\mathrm{G}} \bar{J}_{\mathrm{OMT}} | E_{\mathrm{in}} \rangle,$$

$$\mathscr{P}_{\mathrm{detected}} = \begin{pmatrix} E_x^* & E_y^* \end{pmatrix} \begin{pmatrix} G_1 & 0 \\ 0 & G_2 e^{-i\Delta\phi} \end{pmatrix} \begin{pmatrix} 0 & 1 \\ 1 & 0 \end{pmatrix} \begin{pmatrix} G_1 & 0 \\ 0 & G_2 e^{i\Delta\phi} \end{pmatrix} \begin{pmatrix} E_x \\ E_y \end{pmatrix}.$$

The total power detected at the correlator is a function of the phase difference, $\Delta\phi$,

$$\mathscr{P}_{\mathrm{detected}}(\Delta\phi) = G_1 G_2 \left(E_x^* E_y e^{i\Delta\phi} + E_x E_y^* e^{-i\Delta\phi} \right).$$

In practice, the lock-in measurement will be the difference of the signals in the two states of the phase switch,

$$\mathscr{P}_{\mathrm{lock-in}} = P_{\mathrm{detected}}(0) - \mathscr{P}_{\mathrm{detected}}(\pi) = 2 G_1 G_2 U.$$

5.1.3 Continuous Parameter Modulation Schemes

We conclude our discussion of polarimeter instrumentation by examining more implementation-specific cases. Entries 17–21 in Table 5.2 summarize several such cases. In each example, the polarization modulation parameter is free to take on a continuum of values.

Example 17 illustrates the action of a Faraday rotator. In this case, the modulation can be represented as rotations of the polarization vector in the Q–U plane of the Poincaré sphere. Architectures of this form are the most interesting for astrophysical polarimetry as they allow full characterization of linear polarization. In such implementations, the parameter that is controllable is the phase delay between right and left circular polarization states. Since this is a wavelength-dependent quantity, it is in principle possible to use large sweeps of this parameter as a polarization modulator with arbitrarily large bandwidth. In practice, such factors as achievable magnetic field, drive current magnitude and modulation speed, and thermal dissipation place limits on this type of operation in a typical ferrite-based device. An early implementation of the BICEP polarimeter operating at the South Pole employed Faraday rotators in some of its detection channels (Keating et al. 2003; Moyerman et al. 2013).

Example 18 is similar in concept to the Faraday rotator, although its physical implementation is generally quite different. A variable-delay polarization modulator

[1] Note this is analogous to finding the expectation value of an operator in quantum mechanics.

(VPM) is a device that can insert a variable phase between two linear orthogonal polarization states. For the case in which these states are oriented at an angle of $\pi/4$ with respect to the detector basis, the effect is to rotate the polarization vector on the Poincaré sphere in a plane of constant U. Like the Faraday rotator, the continuous phase delay allows the modulation of polarization signals across large bandwidths. One characteristic of the VPM is that it is insensitive to Stokes U; however, Chuss et al. (2006) describe a technique for placing two such devices in series to fully characterize the polarization state. Such a device was implemented for the Hertz polarimeter at the Submillimeter Telescope Observatory in Arizona (Chuss et al. 2004). Single VPM strategies have been implemented in the Cosmology Large Angular Scale Surveyor (CLASS; Essinger-Hileman et al. 2016) and the Primordial Inflation Polarization Explorer (PIPER; Chuss et al. 2010). These implementations use sky and/or instrument rotation to fully sample the linear polarization in the scene.

Example 19 is perhaps the most commonly-implemented polarization modulation strategy at the wavelengths considered in this book. Rotating half-wave plates offer a similar architecture to the Faraday rotator in Example 17. That is, for the ideal case, they enable modulation of Q and U while leaving Stokes I unchanged. Also, this rotational symmetry prevents systematic mapping between Stokes Q and U. For CMB studies, this leads to minimum ambiguity between E and B modes. Unlike the Faraday rotator, the half-wave plate's variable parameter, its orientation, is not inherently a wavelength-dependent quantity. Thus, historically, these devices have been somewhat bandwidth-limited. However, progress has been made into multiple-layer wave plates with broadband antireflection coatings to increase their effectiveness across the large bandwidths that are commonly used in CMB and magnetic field studies. Examples of half-wave plate-based instruments include far-infrared/submillimeter polarimeters such as: SCU-pol (Greaves et al. 2003), Stokes-KAO (Platt et al. 1991), Hertz-CSO (Schleuning et al. 1997; Dowell et al. 1998), SPARO (Renbarger et al. 2004), SHARP (Novak et al. 2004), BLASTpol (Fissel et al. 2010), HAWC+ (Harper et al. 2018), SCUBA-2 (Bastien et al. 2011), and TolTEC (Lee et al. 2022). As mentioned above, Siringo et al. (2004) have constructed a reflective version of a half-wave plate for use in the submillimeter. CMB experiments that have employed rotating half-wave plates include MAXIPOL (Johnson et al. 2003; EBEX Collaboration et al. 2018), and ABS (Kusaka et al. 2014).

For completeness, we have included the architecture in Example 21, which is an example of a non-ideal polarimeter architecture. This architecture involves a rotating polarizer used in reflection in conjunction with circularly polarized detectors. This architecture has the distinction of being the only example in our compendium that does not have a unitary Jones matrix (half the power is lost). Physically, this is due to the fact that the linear polarizer induces polarization (correlations) in the incoming beam. The sensitivity of the detectors to circular polarization alone allows effective separation between the polarized and unpolarized light. This is a statement that the instrument-induced correlations are not circularly polarized. However, in practice, the detector is sensitive to circular polarization over a narrow finite bandwidth (i.e., outside of this band, Stokes I will be significantly modulated along with Q and U). In summary, many possibilities exist, but not all have equal utility and systematic control in practice.

Example: Direct Detection with Continuous Modulation

We complement the example of a discrete/switched modulation scheme with coherent detectors above with an example of a continuous modulation scheme coupled to direct detectors. For this, we use Example 19 from Table 5.2, a system with a rotating half-wave plate modulator. This is useful for current and future polarimeters that involve MKIDs and TESes, which allow large scalable focal planes to be constructed that dramatically increase mapping speed over early polarimeters. A schematic layout is shown in Figure 5.2,

$$|E_{\text{out}}\rangle = \bar{J}_{\text{HWP}}(\theta)|E_{\text{in}}\rangle. \tag{5.3}$$

We then represent each by their appropriate matrix from Table 5.1. Here we focus on one of the detectors (in this case the one designated by "+"),

$$|E_{\text{out}}\rangle = \begin{pmatrix} \cos 2\theta & \sin 2\theta \\ -\sin 2\theta & \cos 2\theta \end{pmatrix} \begin{pmatrix} E_x \\ E_y \end{pmatrix}.$$

The detected (power) signal from one of the incoherent detectors can be found by multiplying this signal by the linear polarized detector with (power) gain D_+, followed by the complex conjugate of the modulator matrix,

$$\begin{aligned} \mathscr{P}_{\text{detected}}(+) &= \langle E_{\text{in}}|\bar{J}_{\text{HWP}}^{\dagger}\bar{J}_{D_+}\bar{J}_{\text{HWP}}|E_{\text{in}}\rangle \\ &= \begin{pmatrix} E_x^* & E_y^* \end{pmatrix} \begin{pmatrix} \cos 2\theta & -\sin 2\theta \\ \sin 2\theta & \cos 4\theta \end{pmatrix} \begin{pmatrix} D_+ & 0 \\ 0 & 0 \end{pmatrix} \begin{pmatrix} \cos 2\theta & \sin 2\theta \\ -\sin 2\theta & \cos 4\theta \end{pmatrix} \begin{pmatrix} E_x \\ E_y \end{pmatrix}. \end{aligned}$$

The total power detected at the (+) incoherent detector is a function of the half-wave plate angle, θ,

$$\mathscr{P}_{\text{detected}(+)}(\theta) = \frac{D_+}{2}(I + [Q \cos 4\theta + U \sin 4\theta]).$$

The detector that is sensitive to the orthogonal linear polarization (-) would have a similar signal,

$$\mathscr{P}_{\text{detected}(-)}(\theta) = \frac{D_-}{2}(I - [Q \cos 4\theta + U \sin 4\theta]).$$

Note that the difference of these signals can be used to remove the unmodulated Stokes I offset as long as the gains of the two detectors can be calibrated,

$$\mathscr{P}_{\text{detected}(+)}(\theta) - \mathscr{P}_{\text{detected}(-)}(\theta) = \delta(Q \cos 4\theta + U \sin 4\theta).$$

where $\delta \equiv (D_+ + D_-)/2$.

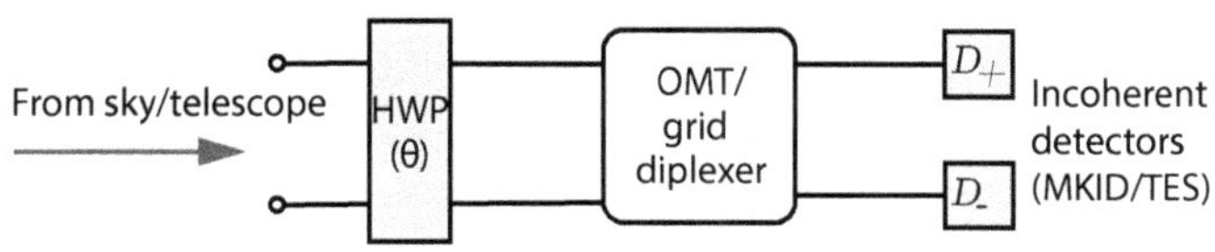

Figure 5.2. Schematic layout for Example 19, an architecture involving direct (incoherent) detection using a with a continuous modulation scheme, in this case, a half-wave plate (HWP).

5.2 A Partial History of Polarimeters: Dust Emission

We present a brief history of polarimetry as focused on those instruments developed primarily for use in Galactic (and extragalactic) dust emission. The history of this field begins with the advent of airborne astronomy from balloons and early aircraft-based observatories such as the Kuiper Airborne Observatory (Cameron 1976). The earliest instrumentation employed to characterize the cosmic microwave background absolute intensity and anisotropy borrowed heavily from the radio and were inherently polarization sensitive. These radiometry experiments placed upper limits on the CMB polarization (e.g., see reported observations by Penzias and Wilson (1965), Nanos (1979), and Smoot and Lubin (1979)). With the detection of the CMB anisotropy in the early 1990s these experimental techniques were redirected toward a targeted search for the polarized signature of inflation.

Though there is some overlap in instrumentation between these two science drivers, for organizational purposes, we separate the compilation of far-infrared through millimeter wave polarimeters between those primarily developed for detection of polarized dust emission and those designed for polarimetry of the CMB; we will discuss CMB polarimetry in more depth in Chapters 6 and 7. Because the significant amount of shared spectrum between Galactic dust emission and the CMB, the distinction between these subfields is artificial and certainly somewhat arbitrary. In this section, we discuss polarimeters operating on single dish telescopes that are designed for pointed observations. This necessarily leaves out both interferometric submillimeter instruments such as ALMA and SMA as well as survey instruments like Planck. The latter was designed as a multi-wavelength CMB mapper, but its 353 GHz band provided a wealth of polarization information crucial to the understanding of magnetic fields in the diffuse ISM as well as tests of dust grain physics. We will discuss Planck's contribution to dust polarimetry in the context of its 353 GHz polarized map of the sky at a diffraction-limited 5′ resolution in Chapter 7.

The polarimeters discussed here are summarized in Table 5.3. We also refer the reader to Pattle & Fissel (2019) for a similar compilation of polarimeters. We have divided the table (and the corresponding discussion) by detector technology categories. As in most measurement-based fields of study, technology drives discovery in the characterization of polarized dust emission. We start with single-pixel detectors enabled by individual semiconducting bolometers or Schottky mixers, depending on the wavelength. Then we discuss advancements achieved from small arrays constructed from tiling such detectors in a focal plane. We follow this with a discussion of microfabricated semiconducting bolometer arrays and conclude with discussions of modern TES arrays that, because of the multiplexing capability, have enabled kilopixel arrays. We finish this chapter with a look ahead to even larger arrays currently being enabled by kinetic inductance detector (KID) technology. These polarimeters that are currently in various stages of development and deployment and are discussed briefly.

5.2.1 Single Pixel Instruments

The first polarimeters in the FIR through millimeter consisted of single pixel semiconducting bolometers, typically cooled via He-3 sorption refrigerators. Early

Table 5.3. A List of Current and Past Single-dish Polarimeters from the Far-infrared Through the Millimeter Part of the EM Spectrum

Instrument	λ (μm)	Ang. Res.	Detector Tech.	Num. Det.	Modulation	References
UCL Polarimeter	77	2′	Bolo.	1	Rotating Grid	Cudlip et al. (1982)
KAO M1	270	90″	Bolo.	1	Sapphire HWP	Hildebrand et al. (1984)
KAO M2	270	90″	Bolo.	1	K-mirror	Dragovan (1986)
KAO 100 μm	100	35″	Bolo.	1	Quartz HWP	Novak et al. (1989)
MILLIPOL	1300	30″	Mixers	1	Grooved Rexolite HWP	Barvainis et al. (1988) Clemens et al. (1990)
FCRAO	1300	25″	Mixers	1	Sapphire HWP	Novak et al. (1990)
Cyclops/SMTO	1300	33″	Bolo.	1	Grooved Rexolite HWP	Glenn et al. (1997)
Aberdeen-QMW	450	18″	Bolo.	1	Quartz HWP	Flett & Murray (1991)
JCMT	800	16″	Bolo.	1	Quartz HWP	Duncan et al. (1990)
	1100	19″	Bolo.	1	Quartz HWP	
Stokes/KAO	100	35″	Bolo.	32	Quartz HWP	Hildebrand et al. (1984)
Hertz/CSO	350	20″	Bolo.	32	Quartz HWP	Dowell et al. (1998) Schleuning (1998)
Hertz/SMTO	850	20″	Bolo.	32	Dual VPM	Krejny et al. (2008) Chuss et al. (2012)
SPARO/Viper	450	4-6′	Bolo.	9	Quartz HWP	Novak et al. (2003) Renbarger et al. (2004)
JCMT/SCU-Pol	450	7.8″	Bolo.	91	Quartz HWP	Holland et al. (1999)
	850	13.8″	Bolo.	37	Quartz HWP	Matthews et al. (2009)
SHARP/CSO	450	9″	Bolo.	144	Quartz HWP	Li et al. (2008)
PolKa/SMTO	1200	20″	Bolo.	19	Grid RHWP	Siringo et al. (2004)
PolKa/APEX	870	20″	Bolo.	295	Grid RHWP	Siringo et al. (2010)
BLASTpol	250	30″	Bolo.	149	Sapphire AHWP	Marsden et al. (2008)
	350	42″	Bolo.	88	Sapphire AHWP	Fissel et al. (2010)
	500	60″	Bolo.	43	Sapphire AHWP	Gandilo et al. (2016)
Planck	850	5′	Bolo.	4	Polarizer Angle	Maffei et al. (2004)
PILOT	240	2.2′	Bolo.	1024	HWP	Misawa et al. (2014)
	550	3.5′	Bolo.	1024	HWP	Mangilli et al. (2019)
JCMT/Pol-2	450	8″	TES	10,000	HWP	Holland et al. (2003)
	850	15″	TES	10,000	HWP	Friberg et al. (2016)
HAWC +/SOFIA	53	4.9″	TES	1280	Quartz HWP	Harper et al. (2018)
	89	7.8″	TES	1280	Quartz HWP	
	154	13.6″	TES	1280	Quartz HWP	
	214	18.2″	TES	1120	Quartz HWP	

In this table "Bolo." refers specifically to a semiconducting bolometer, and "Mixers" refers to Schottky mixers. Horizontal lines are intended to group polarimeters with similar detector technologies.

implementations used gallium-doped germanium and arsenic-doped silicon. These early polarimeters made the first measurements of polarization in a myriad of Galactic molecular clouds. More importantly, these early polarimeters provided platforms to solve the numerous challenges associated with measuring polarimetry, specifically in the area of systematic error control and mitigation of the $1/f^{\alpha}$ noise resulting from the time variations in the system (including the variability of foreground atmospheric and telescope emission). A nice summary of these techniques has been written by Hildebrand et al. (2000).

The initial class of polarimeters in the far-infrared were constructed by adding a rotatable wire grid to the front end of photometers on both NASA's Lear Jet (Dennison et al. 1977) at 85 μm and with the Kuiper Airborne Observatory (Gull et al. 1978, 1980) at similar wavelengths. Both sets of observations targeted the Orion Molecular Cloud and resulted in non-detections. The first detection of far-infrared polarization was done by Cudlip et al. (1982) using Ge–Ga semiconducting bolometer detector technology on a balloon-borne 60 cm telescope. The instrument was sensitive to wavelengths between 40 and 350 μm and also utilized a rotating wire grid to change the state of polarization to which the detectors are sensitive. They reported measurements of $2.2 \pm 0.4\%$ at $16° \pm 8°$ for M42 (Orion) and $0.8 \pm 0.5\%$ at $82° \pm 23°$ for W51A using laboratory measurements of instrumental polarization to calibrate the measurement.

At slightly longer wavelengths, the M1 and M2 polarimeters on the Kuiper Airborne Observatory (KAO) 91 cm telescope operated at 270 μm with an angular resolution of 90′. A free-standing wire grid was used to split the signal into two orthogonal linear polarizations, and each component was detected by a As-doped Si semiconducting bolometers behind Winston non-imaging concentrators (Winston 1970). This instrument made measurements near the peak of the far-infrared emission in Orion at two locations, finding $1.7\% \pm 0.4\%$ at $23° \pm 7°$ and $1.7\% \pm 0.5\%$ at $27° \pm 7°$ (Hildebrand et al. 1984). The instrument was rebuilt to operate at 100 μm (bandpass: 80–135 μm) with an angular resolution of 35″ (Novak et al. 1989). The observing mode for each was to rapidly modulate the secondary mirror of the telescope– to "chop" the detector between the source and a reference position for subsequent demodulation. The beams were switched symmetrically in a process known as "nodding" to compensate for any offsets produced as a result of different illuminations of the primary mirror in the two chop positions of the secondary. This moved the signal band above the $1/f^{\alpha}$ knee of the noise spectrum. Complete polarization coverage was achieved via a rotating sapphire half-wave plate and K-mirror in the M1 and M2 instruments, respectively. The 100 μm version used a quartz half-wave plate.

The James Clerk Maxwell Telescope (JCMT) extended single-detector polarimetry to the submillimeter[2] and millimeter. It hosted the single-pixel Aberdeen-QMW Polarimeter (Flett & Murray 1991), which was used with the UKT14 bolometer

[2] Early papers from the ground refer to this part of the spectrum as "extreme infrared" in addition to "submillimeter." The former term has since mostly fallen out of common use.

camera that operated at 450, 800, and 1100 μm wavelengths (Duncan et al. 1990). This polarimeter employed a stepped quartz half-wave plate, and a similar chop-nod observing mode to the KAO instruments. This instrument detected polarization in a variety of types of sources including cloud cores such as W3, OMC-1, ρ-Oph, Mon R2 (Greaves et al. 1995a), and M17 (Vallee & Bastien 1996). The work from the JCMT also included polarization measurements in pre-T-Tauri and T-Tauri stars (Tamura et al. 1993, 1995b, 1995a); the Galactic center (Greaves et al. 1995b); and measurements of sources with outflows such as S140, GL2591 (Minchin et al. 1996), DCR21, NGC7538-IRS 11 (Minchin & Murray 1994), NGC 1333 (Minchin et al. 1995), and W75N-IRS 1 (Vallee & Bastien 1995).

At milllimeter wavelengths, the MILLIPOL instrument on the NRAO 12 m telescope utilized Schottky mixers along with a grooved rexolite half-wave plate (Clemens et al. 1990). This instrument reported polarization in Orion, with a measurement of 2.9% $\pm$ 0.6% at an angle of 11°, similar to the 77 and 270 μm measurements. A similar measurement was made by Novak et al. (1990) using a 1.3 mm polarimeter on the 14-m telescope at the Five Colleges Radio Astronomy Observatory at a slightly higher resolution. They found a polarization of 2.6% $\pm$ 0.8% at an angle of 40° $\pm$ 8°. Both of these instruments utilized Schottky mixers. The Cyclops polarimeter operated at 1.3 mm using a bolometer system and a similar grooved rexolite HWP as MILLIPOL (Glenn et al. 1997). This instrument measured polarization in DR21 of 1.65% $\pm$ 0.14% at an angle of 16° $\pm$ 2°. They also made the first measurement of polarization in Cepheus A, obtaining an upper limit of 0.9%.

5.2.2 Early Multiple-Detector Polarimetry

The first implementations of arrays in polarimeters utilized multiple copies of single bolometers tiled across the focal plane. An increased number of detectors in the focal plane enabled both an increase in the sensitivity of the observation and the capability to take advantage of different pairings of detectors and sky pixels to mitigate systematic errors. The earliest array polarimeter was the Stokes/KAO polarimeter operated at 100 μm (Platt et al. 1991) with a total of 32 dual-polarization beams on the sky (Platt et al. 1991). This instrument utilized focal planes that consisted of tilings of individual bolometers behind non-imaging concentrators. Similar to the early single-pixel instruments, a free-standing wire grid was used to separate the linear polarizations and direct each component to a separate focal plane. Array polarimetry allowed for faster mapping speeds, enabling maps of such sources as OMC-1 (Schleuning 1998), Sagittarius B2, (Dowell 1997; Novak et al. 1997), and M17 (Dotson 1996). A complete summary of the data taken with the KAO Stokes polarimeter has been published (Dotson et al. 2000). The gain in mapping speed allowed for the geometry of the magnetic field across larger regions of dense clouds. This is an essential step for applying statistical techniques such as the Davis–Chandrasekhar–Fermi method (Davis 1951; Chandrasekhar & Fermi 1953) for statistically determining magnetic field strengths. In addition, array polarimetry allows the measurement of a point on the sky with multiple detectors. This "cross-linking" is helpful in mitigating artifacts in the map making process.

In addition to the KAO, multiple-detector polarimetry was implemented on ground-based polarimeters in the submillimeter. Notable polarimeters include the SCU-Pol polarimeter that used the Submillimeter Common-use Bolometer Array (SCUBA) on the James Clerk Maxwell Telescope (JCMT) in Hawaii (Greaves et al. 2003). SCU-Pol had 91 bolometers in its 450 μm channel and 37 in its 850 μm channel. Due to better observing conditions and control of systematics at the latter, significantly more data were taken at 850 μm. The data repository for this instrument has been published (Matthews et al. 2009). The Hertz polarimeter on the Caltech Submillimeter Observatory (CSO) (Schleuning et al. 1997; Dowell et al. 1998) offered a 32-element polarimetric focal plane at 350 μm at an angular resolution of 20″. The set of observations from the CSO has been collected in Dotson et al. (2010). The Submillimeter Polarimeter for Antarctic Remote Observing (SPARO) (Renbarger et al. 2004) utilized a pair of 9-element bolometer arrays with a half-wave plate on the Viper telescope at the South Pole to discover the toroidal magnetic field in the Galactic center neutral gas layer (Novak et al. 2003) in addition to surveying the large-scale field of star-forming Galactic clouds (Li et al. 2006).

5.2.3 Array Polarimetry-Semiconducting Bolometers

Improved microfabrication techniques led to the production of integrated arrays using semiconducting bolometers. Specifically, the development of the neutron transmutation doped (NTD) process for producing bolometers (Haller et al. 1994) allowed for increased uniformity across thermistors on a single device. Polarimeters have been constructed as front end additions to photometric cameras. For example, the PolKa (*the Polarimeter für Bolometer Kameras*; Siringo et al. 2004) has been developed for NTD bolometer arrays and deployed first at the Heinrich Hertz Telescope on a 19-element camera at a wavelength of 1.2 mm, then later at the APEX telescope on the LABOCA (Kreysa et al. 2003) 295-detector camera. PolKa utilized a rotating reflective half-wave plate and an observing mode in which the primary modulation was done by continuously rotating the wave plate (Siringo et al. 2010, 2012). Their observations include a map of OMC-1 at 870 μm and a measurement of the polarization of Tau-A (Crab Nebula; Wiesemeyer et al. 2014).

The SHARP polarimeter (Novak et al. 2004; Li et al. 2008) utilized a front end optics module that adapted the SHARC-II camera (Voellmer et al. 2003; Dowell et al. 2003) on the Caltech Submillimeter Observatory for polarimetric operation. The module housed a stepped half-wave plate along with beam splitting optics that imaged two orthogonal polarizations on the ends of the 384-detector SHARC-II array, providing a 12 × 12 element polarimetric array at 350 and 450 μm. Observations with SHARP yielded over a dozen papers presenting polarization maps for a range of targets, including isolated low-mass protostellar cores (Davidson et al. 2011; Chapman et al. 2013; Stephens et al. 2013) and Galactic molecular clouds (Houde et al. 2009; Tang et al. 2019).

BLASTpol (Fissel et al. 2010) was a balloon-borne multi-wavelength polarimeter operating in the far-infrared that utilized arrays similar to those used on the Herschel SPIRE instrument (Griffin et al. 2004). BLASTpol observed at 250, 350 and 500 μm with beam sizes of 30, 42 and 60″. BLASTpol observations include the measurement of Lupus I (Matthews et al. 2014) and Vela C (Fissel et al. 2016; Gandilo et al. 2016; Soler et al. 2017; Ashton et al. 2018). An upgraded version of BLASTpol, BLAST-TNG flew 3000 MKID sensors over 3 polarimetric bands in January 2020, serving as a demonstration for future suborbital FIR polarimetry platforms (Coppi et al. 2020).

PILOT (Misawa et al. 2014), another balloon-borne polarimeter, utilized feedhorn-coupled semiconducting bolometer arrays to measure polarization at 240 and 550 μm at arcminute scales. This instrument produced a 240 μm polarimetric map of the large-scale field at the Galactic center, confirming that the magnetic field is oriented roughly along the Galactic plane (Mangilli et al. 2019).

5.2.4 Polarimetry with Transition-Edge Sensor (TES) Arrays

Though they significantly advanced the capability of polarimetry, micromachined arrays of semiconducting (NTD) bolometer are limited in the number of pixels because of the large number of readout lines needed. Transition-edge sensors (TESes) mitigate this challenge through multiplexing. TESes have been implemented in instruments using both time-domain (Irwin 2002) and frequency-domain multiplexing (Lanting et al. 2005) using SQUIDs (Superconducting Quantum Interference Devices).

The POL-2 polarimeter (Friberg et al. 2016) on SCUBA-2 (Holland et al. 2003) implemented a five-layer Pancharatnam sapphire half-wave plate to modulate polarization at 8 Hz for both the 450 and 850 μm channels of SCUBA-2. Operating on the James Clerk Maxwell Telescope, POL-2 has produced a significantly large data set and accompanying scientific advancements, focusing on the B-fields in STar-forming Region Observations (BISTRO) survey (Ward-Thompson et al. 2017). These include results from ρ-Ophiuchus (Pattle et al. 2019; Liu et al. 2019; Perry et al. 2025, Orion Pattle et al. 2017; Lyo et al. 2021; Hwang et al. 2021; Wu et al. 2024), the Galactic center (Lu et al. 2024; Yang et al. 2025), DR21 (Ching et al. 2022), NGC 6334 (Arzoumanian et al. 2021; Tahani et al. 2023), Serpens (Kwon et al. 2022), the “Pillars of Creation” (Pattle et al. 2018), and others (Doi et al. 2020, 2021; Könyves et al. 2021; Ngoc et al. 2021; Yen et al. 2021; Gupta et al. 2022; Hwang et al. 2022; Ward-Thompson et al. 2023; Karoly et al. 2023; Choi et al. 2024; Karoly et al. 2025).

The High Angular-resolution Wideband Camera+ (HAWC+) (Harper et al., 2018) operating on the Stratospheric Observatory For Infrared Astronomy (SOFIA) provided similar capability to POL-2 at far-infrared wavelengths. HAWC+ provided four bands of operation for continuum polarimetry: 53, 89, 154, and 214 μm. A separate set of filters, fore-optics, and wave plate was used at each band such that the detector array Nyquist-sampled the diffraction-limited beam at each wavelength. HAWC+ provided data on multiple clouds, including OMC-1 at all four bands (Chuss et al. 2019; Guerra et al. 2021; Tram et al. 2021; Michail et al. 2021; Lee et al.

2024), GF 9 (Clemens et al. 2018), ρ-Ophiuchus (Santos et al. 2019; Lê et al. 2024), IRAS 15398-3359 (Redaelli et al. 2019), Serpens (Pillai et al. 2020), B335 (Zielinski et al. 2021), G47 (Stephens et al. 2022), Taurus (Li et al. 2022a), M17 (Hoang et al. 2022), OMC-3 & OMC-4 (Li et al. 2022b), 30 Doradus (Tram et al. 2023), and the Flame nebula (Bešlić et al. 2024).

Given its wavelength coverage and sensitivity HAWC+ also enabled measurement of magnetic fields in external Galaxies (Jones et al. 2019, 2020; Marin et al. 2020; Borlaff et al. 2021; Lopez-Rodriguez et al. 2022a, 2022b; Borlaff et al. 2023; Kim et al. 2023), providing a leap forward in understanding the geometry and origin of magnetic fields at the galactic scale. In addition, HAWC+ also surveyed the central 150 pc of our Milky Way (Butterfield et al. 2024; Paré et al. 2024). In Chapter 7, we highlight some of these results in more detail to provide a picture of a current state of the field.

5.2.5 Polarimetrers Using Kinetic Inductance Detector (KID) Arrays

Even with the advantage of the time- and frequency-domain multiplexing strategies, the number of electrical leads running to the cold stage of a detector array remains a limiting factor when evolving beyond kilopixel arrays. Current polarimeters are being designed with kinetic inductance detectors to take advantage of simpler fabrication techniques and more efficient multiplexing.

KIDs enable the readout of multiple detectors using a microwave transmission line, with the backend electronics being standard and relatively straightforward to implement. In each case, the reduction of electrical connections between cryogenic cold stages enable significant factors of multiplexing advantage. There are many experiments currently in development or planned ranging from the far-infrared through to the millimeter. These include NIKA (Ritacco et al. 2017) and NIKA2 (Ajeddig et al. 2022; Ritacco et al. 2022) on the IRAM telescope, TolTEC (Wilson et al. 2020; Lee et al. 2022) on the Large Millimeter Telescope (LMT), the balloon-borne BLAST-TNG (Coppi et al. 2020), and CCAT-Prime on the Fred Young Submillimeter telescope (FYST; CCAT-Prime Collaboration et al. 2023).

In addition to these, as of this writing, the PRIMA (Burgarella et al. 2024) mission is in a Phase A study as a NASA probe mission. This cryogenic telescope will contain the PRIMAger camera, which will include polarimetric imaging from 80 to 260 μm. The low backgrounds and good systematic control that will be provided from a cryogenic telescope in space will provide unprecedented understanding of magnetic fields and dust grain physics.

5.2.6 Interferometry

Our focus in this book is polarimetry using single-dish telescopes. The subject of instrumentation using interferometric techniques is sufficiently complex for its own separate volume (Thompson et al. 2017), and indeed its use in high-resolution polarimetry is crucial for understanding the role of magnetic fields in astrophysical settings. High resolution submillimeter and millimeter interferometric observations have, however, already produced significant discoveries. We discuss these briefly in Chapter 7.

References

Ajeddig, H., Adam, R., Ade, P., et al. 2022, Conf. Ser. 257, mm Universe @ NIKA2—Observing the mm Universe with the NIKA2 Camera 00002 eds. M. De Petris, A. Ferragamo and F. Mayet (Les Ulis: EDP Sciences)
Arzoumanian, D., Furuya, R. S., Hasegawa, T., et al. 2021, A&A, 647, A78
Ashton, P. C., Ade, P. A. R., Angilè, F. E., et al. 2018, ApJ, 857, 10
Barvainis, R., Clemens, D. P., & Leach, R. 1988, AJ, 95, 510
Bastien, P., Bissonnette, E., Simon, A., et al. 2011, Conf. Ser. 449, Astronomical Polarimetry 2008: Science from Small to Large Telescopes ed. P. Bastien, N. Manset, D. P. Clemens, & N. St-Louis (San Francisco, CA: ASP) 68
Bešlić, I., Coudé, S., Lis, D. C., et al. 2024, A&A, 684, A212
Bischoff, C., Brizius, A., Buder, I., et al. 2013, ApJ, 768, 9
Borlaff, A. S., Lopez-Rodriguez, E., Beck, R., et al. 2023, ApJ, 952, 4
Borlaff, A. S., Lopez-Rodriguez, E., Beck, R., et al. 2021, ApJ, 921, 128
Burgarella, D., Ciesla, L., Sauvage, M., et al. 2024, Proc. SPIE, 13092, 130923B
Butterfield, N. O., Guerra, J. A., Chuss, D. T., et al. 2024, ApJ, 968, 63
Cameron, R. M. 1976, S&T, 52, 327
CCAT-Prime CollaborationAravena, M., Austermann, J. E., Basu, K., et al. 2023, ApJS, 264, 7
Chandrasekhar, S., & Fermi, E. 1953, ApJ, 118, 113
Chapman, N. L., Davidson, J. A., Goldsmith, P. F., et al. 2013, ApJ, 770, 151
Ching, T.-C., Qiu, K., Li, D., et al. 2022, ApJ, 941, 122
Choi, Y., Kwon, W., Pattle, K., et al. 2024, ApJ, 977, 32
Chuss, D. T., Ade, P. A. R., Benford, D. J., et al. 2010, Proc. SPIE, 7741, 77411P
Chuss, D. T., Andersson, B. G., Bally, J., et al. 2019, ApJ, 872, 187
Chuss, D. T., Benford, D. J., Walker, C., et al. 2004, Proc. SPIE, 5492, 1450
Chuss, D. T., Wollack, E. J., Henry, R., et al. 2012, ApOpt, 51, 197
Chuss, D. T., Wollack, E. J., Moseley, S. H., & Novak, G. 2006, ApOpt, 45, 5107
Clemens, D. P., El-Batal, A. M., Cerny, C., et al. 2018, ApJ, 867, 79
Clemens, D. P., Leach, R. W., Barvainis, R., & Kane, B. D. 1990, PASP, 102, 1064
Coppi, G., Ade, P. A. R., Ashton, P. C., et al. 2020, Proc. SPIE, 11445, 1144526
Cudlip, W., Furniss, I., King, K. J., & Jennings, R. E. 1982, MNRAS, 200, 1169
Davidson, J. A., Novak, G., Matthews, T. G., et al. 2011, ApJ, 732, 97
Davis, L. 1951, PhRv, 81, 890
Dennison, B., Ward, D. B., Gull, G. E., & Harwit, M. 1977, AJ, 82, 39
Doi, Y., Hasegawa, T., Furuya, R. S., et al. 2020, ApJ, 899, 28
Doi, Y., Tomisaka, K., Hasegawa, T., et al. 2021, ApJL, 923, L9
Dotson, J. L. 1996, ApJ, 470, 566
Dotson, J. L., Davidson, J., Dowell, C. D., Schleuning, D. A., & Hildebrand, R. H. 2000, ApJS, 128, 335
Dotson, J. L., Vaillancourt, J. E., Kirby, L., et al. 2010, ApJS, 186, 406
Dowell, C. D. 1997, ApJ, 487, 237
Dowell, C. D., Allen, C. A., Babu, R. S., et al. 2003, Proc. SPIE, 4855, 73
Dowell, C. D., Hildebrand, R. H., Schleuning, D. A., et al. 1998, ApJ, 504, 588
Dragovan, M. 1986, ApJ, 308, 270
Duncan, W. D., Robson, E. I., Ade, P. A. R., Sandell, G., & Griffin, M. J. 1990, MNRAS, 243, 126

EBEX CollaborationAboobaker, A. M., Ade, P., Araujo, D., et al. 2018, ApJS, 239, 7
Essinger-Hileman, T., Kusaka, A., Appel, J. W., et al. 2016, RScI, 87, 094503
Fissel, L. M., Ade, P. A. R., Angilè, F. E., et al. 2016, ApJ, 824, 134
Fissel, L. M., Ade, P. A. R., Angilè, F. E., et al. 2010, Proc. SPIE, 7741, 77410E
Flett, A. M., & Murray, A. G. 1991, MNRAS, 249, 4P
Friberg, P., Bastien, P., Berry, D., et al. 2016, Proc. SPIE, 9914, 991403
Gandilo, N. N., Ade, P. A. R., Angilè, F. E., et al. 2016, ApJ, 824, 84
Glenn, J., Walker, C. K., & Young, E. T. 1997, IJIMW, 18, 285
Greaves, J. S., Holland, W. S., Jenness, T., et al. 2003, MNRAS, 340, 353
Greaves, J. S., Holland, W. S., & Murray, A. G. 1995a, A&A, 297, L49
Greaves, J. S., Holland, W. S., Murray, A. G., & Nartallo, R. 1995b, MNRAS, 272, L1
Griffin, M. J., Swinyard, B. M., & Vigroux, L. 2004, Proc. SPIE, 5487, 413
Guerra, J. A., Chuss, D. T., Dowell, C. D., et al. 2021, ApJ, 908, 98
Gull, G. E., Houck, J. R., McCarthy, J. F., Forrest, W. J., & Harwit, M. 1978, AJ, 83, 1440
Gull, G. E., Russell, R. W., Melnick, G., & Harwit, M. 1980, AJ, 85, 1379
Gupta, A., Yen, H.-W., Koch, P., et al. 2022, ApJ, 930, 67
Haller, E. E., Itoh, K. M., Beeman, J. W., Hansen, W. L., & Ozhogin, V. I. 1994, Proc. SPIE, 2198, 630
Harper, D. A., Runyan, M. C., Dowell, C. D., et al. 2018, JAI, 7, 1840008
Hildebrand, R. H., Davidson, J. A., Dotson, J. L., et al. 2000, PASP, 112, 1215
Hildebrand, R. H., Dragovan, M., & Novak, G. 1984, ApJL, 284, L51
Hoang, T. D., Ngoc, N. B., Diep, P. N., et al. 2022, ApJ, 929, 27
Holland, W. S., Duncan, W., Kelly, B. D., et al. 2003, Proc. SPIE, 4855, 1
Holland, W. S., Robson, E. I., Gear, W. K., et al. 1999, MNRAS, 303, 659
Houde, M., Vaillancourt, J. E., Hildebrand, R. H., Chitsazzadeh, S., & Kirby, L. 2009, ApJ, 706, 1504
Hwang, J., Kim, J., Pattle, K., et al. 2021, ApJ, 913, 85
Hwang, J., Kim, J., Pattle, K., et al. 2022, ApJ, 941, 51
Irwin, K. D. 2002, PhyC, 368, 203
Jarosik, N., Barnes, C., Bennett, C. L., et al. 2003a, ApJS, 148, 29
Jarosik, N., Bennett, C. L., Halpern, M., et al. 2003b, ApJS, 145, 413
Johnson, B. R., Abroe, M. E., Ade, P., et al. 2003, NewAR, 47, 1067
Jones, T. J., Dowell, C. D., Lopez Rodriguez, E., et al. 2019, ApJL, 870, L9
Jones, T. J., Kim, J.-A., Dowell, C. D., et al. 2020, AJ, 160, 167
Karoly, J., Ward-Thompson, D., Pattle, K., et al. 2023, ApJ, 952, 29
Karoly, J., Ward-Thompson, D., Pattle, K., et al. 2025, ApJL, 982, L22
Keating, B. G., Ade, P. A. R., Bock, J. J., et al. 2003, Proc. SPIE, 4843, 284
Kim, J.-A., Jones, T. J., & Dowell, C. D. 2023, AJ, 165, 223
Könyves, V., Ward-Thompson, D., Pattle, K., et al. 2021, ApJ, 913, 57
Krejny, M., Chuss, D., D'Aubigny, C. D., et al. 2008, ApOpt, 47, 4429
Kreysa, E., Bertoldi, F., Gemuend, H.-P., et al. 2003, Proc. SPIE, 4855, 41
Kusaka, A., Essinger-Hileman, T., Appel, J. W., et al. 2014, RScI, 85, 039901
Kwon, W., Pattle, K., Sadavoy, S., et al. 2022, ApJ, 926, 163
Lanting, T. M., Cho, H.-M., Clarke, J., et al. 2005, ApPhL, 86, 112511
Lê, N., Tram, L. N., Karska, A., et al. 2024, A&A, 690, A191
Lee, D., Chen, C.-Y., Novak, G., et al. 2024, ApJ, 972, 102

Lee, D., Novak, G., Berthoud, M., et al. 2022, Proc. SPIE, 12190, 121901O
Li, H., Dowell, C. D., Kirby, L., Novak, G., & Vaillancourt, J. E. 2008, ApOpt, 47, 422
Li, H., Griffin, G. S., Krejny, M., et al. 2006, ApJ, 648, 340
Li, P. S., Lopez-Rodriguez, E., Ajeddig, H., et al. 2022a, MNRAS, 510, 6085
Li, P. S., Lopez-Rodriguez, E., Soam, A., & Klein, R. I. 2022b, MNRAS, 514, 3024
Liu, J., Qiu, K., Berry, D., et al. 2019, ApJ, 877, 43
Lopez-Rodriguez, E., Clarke, M., Shenoy, S., et al. 2022a, ApJ, 936, 65
Lopez-Rodriguez, E., Mao, S. A., Beck, R., et al. 2022b, ApJ, 936, 92
Lu, X., Liu, J., Pillai, T., et al. 2024, ApJ, 962, 39
Lubin, P. M., & Smoot, G. F. 1979, PhRvL, 42, 129
Lyo, A. R., Kim, J., Sadavoy, S., et al. 2021, ApJ, 918, 85
Maffei, B., Ade, P. A., Bock, J. J., et al. 2004, Proc. SPIE, 5487, 523
Mangilli, A., Aumont, J., Bernard, J. P., et al. 2019, A&A, 630, A74
Marin, F., Le Cam, J., Lopez-Rodriguez, E., et al. 2020, MNRAS, 496, 215
Marsden, G., Ade, P. A. R., Benton, S., et al. 2008, Proc. SPIE, 702002, 702002
Matthews, B. C., McPhee, C. A., Fissel, L. M., & Curran, R. L. 2009, ApJS, 182, 143
Matthews, T. G., Ade, P. A. R., Angilè, F. E., et al. 2014, ApJ, 784, 116
Michail, J. M., Ashton, P. C., Berthoud, M. G., et al. 2021, ApJ, 907, 46
Minchin, N. R., Bonifacio, V. H. R., & Murray, A. G. 1996, A&A, 315, L5
Minchin, N. R., & Murray, A. G. 1994, A&A, 286, 579
Minchin, N. R., Sandell, G., & Murray, A. G. 1995, A&A, 293, L61
Misawa, R., Bernard, J. P., Ade, P., et al. 2014, Proc. SPIE, 9153, 91531
Moyerman, S., Bierman, E., Ade, P. A. R., et al. 2013, ApJ, 765, 64
Nanos, Jr., G. P. 1979, ApJ, 232, 341
Ngoc, N. B., Diep, P. N., Parsons, H., et al. 2021, ApJ, 908, 10
Novak, G., Chuss, D. T., Davidson, J. A., et al. 2004, Proc. SPIE, 5498, 278
Novak, G., Chuss, D. T., Renbarger, T., et al. 2003, ApJL, 583, L83
Novak, G., Dotson, J. L., Dowell, C. D., et al. 1997, ApJ, 487, 320
Novak, G., Gonatas, D. P., Hildebrand, R. H., & Platt, S. R. 1989, PASP, 101, 215
Novak, G., Predmore, C. R., & Goldsmith, P. F. 1990, ApJ, 355, 166
Paré, D., Butterfield, N. O., Chuss, D. T., et al. 2024, ApJ, 969, 150
Pattle, K., & Fissel, L. 2019, FrASS, 6, 15
Pattle, K., Lai, S.-P., Hasegawa, T., et al. 2019, ApJ, 880, 27
Pattle, K., Ward-Thompson, D., Berry, D., et al. 2017, ApJ, 846, 122
Pattle, K., Ward-Thompson, D., Hasegawa, T., et al. 2018, ApJL, 860, L6
Penzias, A. A., & Wilson, R. W. 1965, ApJ, 142, 419
Perry, J. P., Pattle, K., Johnstone, D., et al. 2025, MNRAS, 536, 1736
Pillai, T. G. S., Clemens, D. P., Reissl, S., et al. 2020, NatAs, 4, 1195
Platt, S. R., Hildebrand, R. H., Pernic, R. J., Davidson, J. A., & Novak, G. 1991, PASP, 103, 1193
Redaelli, E., Alves, F. O., Santos, F. P., & Caselli, P. 2019, A&A, 631, A154
Renbarger, T., Chuss, D. T., Dotson, J. L., et al. 2004, PASP, 116, 415
Ritacco, A., Adam, R., Ade, P., et al. 2022, Conf. Ser. 228, mm Universe @ NIKA2—Observing the mm Universe with the NIKA2 Camera, 00022, eds. M. De Petris, A. Ferragamo and F. Mayet (Les Ulis: EDP Sciences) 00021
Ritacco, A., Ponthieu, N., Catalano, A., et al. 2017, A&A, 599, A34

Santos, F. P., Chuss, D. T., Dowell, C. D., et al. 2019, ApJ, 882, 113
Schleuning, D. A. 1998, ApJ, 493, 811
Schleuning, D. A., Dowell, C. D., Hildebrand, R. H., & Platt, S. R. 1997, PASP, 109, 307
Siringo, G., Kovács, A., Kreysa, E., et al. 2012, Proc. SPIE, 8452, 845206
Siringo, G., Kreysa, E., Kovács, A., Menten, K. M., & Forbrich, J. 2010, Proc. SPIE, 7741, 774108
Siringo, G., Kreysa, E., Reichertz, L. A., & Menten, K. M. 2004, A&A, 422, 751
Soler, J. D., Ade, P. A. R., Angilè, F. E., et al. 2017, A&A, 603, A64
Stephens, I. W., Looney, L. W., Kwon, W., et al. 2013, ApJL, 769, L15
Stephens, I. W., Myers, P. C., Zucker, C., et al. 2022, ApJL, 926, L6
Tahani, M., Bastien, P., Furuya, R. S., et al. 2023, ApJ, 944, 139
Tamura, M., Hayashi, S., & Hough, J. H. 1995a, Ap&SS, 224, 21
Tamura, M., Hayashi, S. S., Yamashita, T., Duncan, W. D., & Hough, J. H. 1993, ApJL, 404, L21
Tamura, M., Hough, J. H., & Hayashi, S. S. 1995b, ApJ, 448, 346
Tang, Y.-W., Koch, P. M., Peretto, N., et al. 2019, ApJ, 878, 10
Thompson, A. R., Moran, J. M., & Swenson, G. W. 2017, Interferometry and Synthesis in Radio Astronomy (New York: Springer)
Tram, L. N., Bonne, L., Hu, Y., et al. 2023, ApJ, 946, 8
Tram, L. N., Lee, H., Hoang, T., et al. 2021, ApJ, 908, 159
Vallee, J. P., & Bastien, P. 1995, A&A, 294, 831
Vallee, J. P., & Bastien, P. 1996, A&A, 313, 255
Voellmer, G. M., Allen, C. A., Amato, M. J., et al. 2003, Proc. SPIE, 4855, 63
Ward-Thompson, D., Karoly, J., Pattle, K., et al. 2023, ApJ, 946, 62
Ward-Thompson, D., Pattle, K., Bastien, P., et al. 2017, ApJ, 842, 66
Wiesemeyer, H., Hezareh, T., Kreysa, E., et al. 2014, PASP, 126, 1027
Wilson, G. W., Abi-Saad, S., Ade, P., et al. 2020, Proc. SPIE, 11453, 1145302
Winston, R. 1970, JOSA, 6, 245
Wu, J., Qiu, K., Poidevin, F., et al. 2024, ApJL, 977, L31
Yang, M.-Z., Lai, S.-P., Karoly, J., et al. 2025, ApJ, 983, 184
Yen, H.-W., Koch, P. M., Hull, C. L. H., et al. 2021, ApJ, 907, 33
Zielinski, N., Wolf, S., & Brunngräber, R. 2021, A&A, 645, A125

Chapter 6

Observations and Data Reduction

Measuring astrophysical polarization is in general challenging, given that source polarization fluxes are generally significantly smaller than those of the total intensity. This is true for both the measurement of polarized dust emission and for that of the cosmic microwave background. In each of these cases, the technical challenge stems not just from the relatively small signal size but also in the need to measure these signals in the presence of large, and often fluctuating, backgrounds. In the language of Chapter 2, these backgrounds are non-ergodic and thus have an indeterminate mean. These backgrounds can be from the sources themselves, but often the more problematic background contamination stems from thermal emission from the atmosphere (when not space-based) and the instrument itself.

There is some good news. The polarized nature of the desired signals provide additional degrees of freedom that can be utilized to distinguish the signal of interest from fluctuating backgrounds and noise. Polarimeters must of course have sensitivity to polarization, but this is not sufficient to fully measure the polarization of the astrophysical source. Successful implementation of polarimetry requires some sort of polarization modulation. That is, systematic variation in the polarization sensitivity of the instrument that can be subsequently demodulated in a way to null or isolate the unwanted signals. Polarimeters do this either using an explicit polarization modulator (e.g., a half-wave plate; see Chapter 4), through the telescope scanning strategy, or both. These spatial-temporal encoding and demodulation processes can also provide paths to signal stabilization and image calibration. Astronomical polarimetry measurements involve multiple observations between which instrument parameters are changed. These parameters include the modulator state, telescope pointing, sky/instrument rotation, etc. The results of the observations is a *timestream* as a data product, that when linked to the telemetry of the instrument can directly test astrophysical hypotheses.

Timestreams contain all of the information needed to address a given scientific question, but they are highly redundant and not intuitive for the process of interpretation. Thus, it is desirable to *reduce* these into images or maps. Maps provide

doi:10.1088/2514-3433/ae1022ch6

a data product that is complete, but as compact as possible (Wright et al. 1996). A map serves as the starting point for data analysis and interpretation. For polarimetry measurements, maps of Q and U in addition to total intensity (Stokes I) are required. For CMB measurements, maps can be turned into *angular power spectra*, which are the key data products for constraining cosmological parameters and are constructed from the maps after foregrounds are removed.

A map (or image) is a filtered or approximate representation of the scene as described in Chapter 4. The amount of information captured is dependent on the detailed instrument configuration, sampling (spectral, spatial, temporal, etc.), as well as the strategy for observing the scene. Instruments record a finite amount of information set by a variety of factors including sky coverage, angular resolution, and calibration (which can be linked to scan strategy). In addition to their effect on the map itself, these limits are manifest as structure in the associated spatial spectrum. For example, the angular resolution limit corresponds to a high frequency cut-off in the spectrum; the map extent limits its low frequency support. The scan strategy, which can be judiciously designed based on consideration of the problem, links the time domain quantities of the timestream to spatial quantities in the map domain.

6.1 Map-Making

Generally, data products include maps of the Stokes parameters I, Q, U, and V (if relevant) and their associated uncertainties, including covariance.

Astronomical polarization, as we have discussed, is a small fraction of a larger, unpolarized signal. For the case of Galactic dust, this signal is ~1%–20% of the unpolarized thermal emission. For the case of the CMB, it may be as low as 10^{-9} of the total CMB intensity. In considering the process of making maps from time-ordered data, it is useful to articulate the content of a given timestream. Typical elements include:

- The signal of interest from the scene, in all relevant Stokes parameters.
- White (uncorrelated) noise from the detector, source, and foregrounds/backgrounds, including potentially telescope and atmospheric emission.
- Correlated noise from varying backgrounds ($1/f$ noise).
- Changes in instrument gain.
- Transient signals (e.g., glitches, spikes, electromagnetic interference, ionizing radiation, etc.).
- Systematic signals (e.g., ground pickup, instrumental polarization signatures, etc.).

Ultimately, each instrument has its unique requirements and thus the level at which each of these elements are present varies significantly. For example, a space-borne cold telescope like the proposed PRIMA mission (Burgarella et al. 2024) will have significantly different levels of $1/f$ noise, systematic signals, and backgrounds than a ground-based instrument in the Atacama Desert. In addition, the required fidelity for reconstructing the maps varies according to the science case. Systematics will vary between instrument systems and polarimeter optical configurations as well.

6.1.1 Observing Strategy

Key to successful measurements, the ability to effectively remove correlated noise can be improved significantly by choosing the observing strategy appropriately. Making repeated measurements of the sky with the instrument in a similar state before the level of correlated noise changes appreciably moves the signal band away from the $1/f$-dominated part of the spectrum. This can be done by repeating the measurement of a sky pixel within this timescale, by rapidly modulating the polarization sensitivity of the instrument, or some combination of these mitigation strategies (Miller et al., 2016).

Early implementations of observing strategies for FIR through millimeter polarimetry utilized rapid position switching (or "chopping") of the telescope beam between two or more sky positions. This and its associated techniques were borrowed from imaging experiments, first in the radio and infrared, then later in the FIR. Early strategies performed the differencing in hardware via a lock-in amplifier and stored the demodulated data. The difference of these signals removes the common background intensity, which is approximately constant over this short timescale. Hildebrand et al. (2000) details a version of this technique for early FIR and submillimeter polarimetry where the observations using a two-point chop (nominally one "on-source" and one "reference") were symmetrized by "nodding," which interchanged the roles of the two beams in the chop during the observing sequence. This accounted for, among other effects, offsets due to differential illumination of the primary mirror when utilizing motion of the secondary to direct the beam.

Increased computing and storage capabilities led to the ability to store the entirety of the discretely sampled timestream, leaving the demodulation to the reduction software. This allowed for more flexible data acquisition and reduction processes for focal plane arrays. The resulting observation strategies have entailed scanning the detector array across the sky repeatedly, making sure to return to the same point in the sky regularly to set the timescale for correlated noise mitigation. This strategy has been called "on-the-fly" or "scan-mode" mapping (Emerson et al. 1979; Mangum et al. 2007). Scanning strategies that maximize cross-linking between instrument beams and map pixels (such as Lissajous) are beneficial for recovering large-scale modes and for improved noise performance (Wright et al. 1996; Tegmark 1997; Kovács 2008b). Still other observing strategies use a fast polarization modulation in concert with a slower scan (Kusaka et al. 2014; Harrington et al. 2021).

6.1.2 Formalism

A common formalism that relates the timestream to the final map data products is described in multiple places in the literature (Tegmark 1997; Sutton et al. 2010). The timestream is represented by a vector $\mathbf{y}$ is related to the map, which can also be represented by a vector containing the values of the map pixels, by the data model,

$$\mathbf{y} = \mathbf{A}\mathbf{x} + \mathbf{n}. \tag{6.1}$$

Here, $\mathbf{A}$ is called the *pointing matrix* (Tegmark 1997) or more generally, the *observing matrix* (Kovács 2008b), since it can also include information about the instrument's polarization state and potentially the beam shape. The function of $\mathbf{A}$ is to pair a given observation to a pixel on the sky, taking into account the relevant instrument parameters. The form of $\mathbf{A}$ reflects the observing strategy. The vector $\mathbf{n}$ is a vector representing the noise.

The length of $\mathbf{y}$ corresponds to the total number of data samples acquired from the output of the detectors, which we will define as N. The length of $\mathbf{x}$ is the total number of map pixels, M. $\mathbf{A}$ is then an $M \times N$ sparse matrix. For polarimetry, it is common to include all relevant Stokes maps in $\mathbf{x}$. The $\mathbf{A}$ matrix then includes not only information about which pixels are connected to a given pointing, but also instrument rotation and (if applicable) polarization modulator state.

Once a problem can be couched in this formalism, a solution is straightforward to implement as a linear least-squares minimization for the case of noiseless data (or equivalently, uniformly-weighted white noise). A measurement of the scene is given by the estimator $\hat{\mathbf{x}}$:

$$\hat{\mathbf{x}} = (\mathbf{A}^T\mathbf{N}^{-1}\mathbf{A})^{-1}\mathbf{A}^T\mathbf{N}^{-1}\mathbf{y}. \tag{6.2}$$

Here, $\mathbf{N} = \langle \mathbf{n}\mathbf{n}^T \rangle$ as the covariance matrix. If there is no correlated noise in the system, the solution is reduced to

$$\hat{\mathbf{x}} = (\mathbf{A}^T\mathbf{A})^{-1}\mathbf{A}^T\mathbf{y}. \tag{6.3}$$

The direct matrix inversion can generate an exact solution; however, this becomes computationally prohibitive for most modern data sets given their size. There are approaches to address this computational challenge, involving mitigating the correlated noise in the timestreams before the map-making step. One can model the noise components by fitting the time stream with slowly varying polynomials or utilizing common-mode responses across detectors in the focal plane. This process, known as *destriping*, provides an estimate of the correlated noise. This is then subtracted from the time stream (Tristram et al. 2011). *Pre-whitening*, which involves the filtering of the timestream to remove correlations (at the expense of raising the white noise level) can also be used. After this preprocessing, maps are made from the decorrelated data set. The data model can be solved for an estimate of the map either via an iterative minimization (e.g., preconditioned conjugate gradient (Li et al. 2023) method) or through direct co-adding of the observations of each pixel.

There are many such software frameworks that have been developed to accompany the myriad of polarimeters. An example of such a software product that is publicly available is CRUSH (Kovács 2008a).

6.2 Change of Basis: $\{q, u\} \rightarrow \{p, \phi\}$

In applications where one measures interstellar magnetic fields, one is generally concerned with the polarization at a particular point in space. Although measurements are almost always made using Stokes parameters, one is often interested in the polarization fraction, p, and the polarization angle, ϕ. Measurements are typically made in the $\{q, u\}$ basis (derived from maps of Q, U, and I), whereas interpretation is often done in the $\{p, \phi\}$ basis.

The transformations in the desired quantities are

$$p = \sqrt{q^2 + u^2}, \tag{6.4}$$

$$\phi = \frac{1}{2}\arctan\frac{u}{q}. \tag{6.5}$$

Here, we have used Equation (2.30) and assumed $v = 0$. The errors in p and ϕ can be found from standard error propagation (Bevington & Robinson 1969). This leads to the following expression for the uncertainty in p,

$$\sigma_p^2 = \left(\frac{\partial p}{\partial q}\right)^2 \sigma_q^2 + \left(\frac{\partial p}{\partial u}\right)^2 \sigma_u^2 + \left(\frac{\partial p}{\partial q}\right)\left(\frac{\partial p}{\partial u}\right)\sigma_{qu}^2 = \frac{1}{p^2}\left(q^2\sigma_q^2 + u^2\sigma_u^2 + qu\sigma_{qu}^2\right). \tag{6.6}$$

Here, σ_q^2, σ_u^2, and σ_{qu}^2 are the variance in q, variance in u, and the covariance between q and u, respectively. In many applications, one assumes q and u are independent and this expression reduces to

$$\sigma_p^2 = \left(\frac{\partial p}{\partial q}\right)^2 \sigma_q^2 + \left(\frac{\partial p}{\partial u}\right)^2 \sigma_u^2 = \frac{1}{p^2}\left(q^2\sigma_q^2 + u^2\sigma_u^2\right). \tag{6.7}$$

Because q and u are related by a rotation, and because a good measurement strategy treats them equivalently, it is useful to consider this expression for the case where the errors in q and u are equal ($\sigma_q = \sigma_u$). For this case,

$$\sigma_p = \sigma_q. \tag{6.8}$$

For the polarization angle, one finds,

$$\sigma_\phi^2 = \left(\frac{\partial \phi}{\partial q}\right)^2 \sigma_q^2 + \left(\frac{\partial \phi}{\partial u}\right)^2 \sigma_u^2 + \left(\frac{\partial \phi}{\partial q}\right)\left(\frac{\partial \phi}{\partial u}\right)\sigma_{qu}^2 = \frac{1}{4p^4}\left(u^2\sigma_q^2 + q^2\sigma_u^2 - qu\sigma_{qu}^2\right). \tag{6.9}$$

If we assume q and u to be independent,

$$\sigma_\phi^2 = \left(\frac{\partial \phi}{\partial q}\right)^2 \sigma_q^2 + \left(\frac{\partial \phi}{\partial u}\right)^2 \sigma_u^2 = \frac{1}{4p^4}\left(u^2\sigma_q^2 + q^2\sigma_u^2\right). \tag{6.10}$$

Again, if $\sigma_q = \sigma_u$, one finds,

$$\sigma_\phi = \frac{\sigma_q}{2p} = \frac{\sigma_p}{2p}. \tag{6.11}$$

We can define the signal-to-noise ratio of the polarization fraction to be $\mathrm{SNR} \equiv p/\sigma_p$. Under these assumptions, the uncertainty in the polarization angle is only a function of SNR,

$$\sigma_\phi = \frac{1}{2\,(\mathrm{SNR})} = \frac{28.6^\circ}{\mathrm{SNR}}. \tag{6.12}$$

This justifies a common rule-of-thumb that for SNR=3, the uncertainty of the polarization angle is $\sim 10^\circ$. For studies of magnetic fields in the ISM, this is a standard threshold for reporting polarization vectors, though in some instances, authors have justified reporting results at lower SNR.

6.3 Debiasing

Unlike q and u, which are signed quantities that can obey Gaussian statistics, p is a positive-definite quantity, given its definition in Equation (6.4). Because of this, any noise present will bias p toward larger values. Thus, values of p, especially under the condition of low signal-to-noise, should be debiased before interpretation.

6.3.1 Bayesian Approach

This general problem has been treated by Simmons & Stewart (1985) and Vaillancourt (2006). Here, we treat the problem using a Bayesian approach similar to that of the latter. Ultimately, we would like to know the posterior conditional probability, $P(p_0|p)$, the probability that the true polarization is p_{0_o}, given a measured polarization p. The probability distributions are only dependent on the polarization fraction signal-to-noise, so we follow Simmons & Stewart (1985) and scale the polarization by the uncertainty, σ_p. That is, p and p_0 really represent p/σ_p and p_0/σ_p, respectively. This is consistent with the work of Vaillancourt (2006) if one sets $\sigma = 1$. We start with Bayes' Theorem, appropriately tuned to the problem at hand,

$$P(p_0|p)P(p) = P(p|p_0)P(p_0) \tag{6.13}$$

$$P(p_0|p) = \frac{P(p|p_0)P(p_0)}{P(p)}. \tag{6.14}$$

Here, $P(p)$ is the probability of measuring a polarization p for any source polarization. $P(p|p_0)$ is the likelihood function, the probability of measuring p given a true polarization of p_0. Finally, $P(p_0)$ is the probability that the true polarization is p_0. This is the prior. The quantity $P(p)$ is the probability for measuring p for any value of p_0. This is just the integral of the likelihood,

$$P(p) = \int_0^1 P(p|p_0)dp_0. \tag{6.15}$$

Thus, the key to understanding this is to calculate the likelihood. We can do this by starting with the Stokes parameters. We calculate the probability of measuring Stokes parameters $\{q, u\}$ given a true set of Stokes parameters $\{q_0, u_0\}$,

$P(\{q, u\}|\{q_0, u_0\})$. This joint probability distribution is just the product of those for the individual Stokes parameters, which are Gaussian,

$$P(\{q, u\}|\{q_0, u_0\}) = \left(\frac{1}{\sqrt{2\pi}}\right)^2 \exp\left\{-\frac{(q - q_0)^2}{2} - \frac{(u - u_0)^2}{2}\right\}. \tag{6.16}$$

This assumes that $\sigma \equiv \sigma_q = \sigma_u = 1$. Simplifying we find,

$$P(\{q, u\}|\{q_0, u_0\}) = \frac{1}{2\pi} \exp\left\{-\frac{q^2 + u^2 + q_0^2 + u_0^2 - 2qq_0 - 2uu_0}{2}\right\}. \tag{6.17}$$

We can now make the substitutions described in Equations (6.4) and (6.5) to find the joint probability distribution in terms of p, ϕ as follows:

$$P(\{p, \phi\}|\{p_0, \phi_0\}) = \frac{1}{2\pi} \exp\left\{-\frac{p^2 + p_0^2 - 2(p\cos 2\phi p_0 \cos 2\phi_0 + p \sin 2\phi p_0 \sin 2\phi_0)}{2}\right\} \tag{6.18}$$

$$= \frac{1}{2\pi} \exp\left\{-\frac{p^2 + p_0^2 - 2pp_0 \cos 2(\phi - \phi_0)}{2}\right\}. \tag{6.19}$$

This is the joint probability function for measuring $\{p, \phi\}$, given true values of $\{p_0, \phi_0\}$. We are concerned with $P(p|p_0)$, the probability of measuring p given p_0, regardless of the angle of polarization. To do this, we integrate the joint probability distribution over $\phi - \phi_0$ to marginalize over all possible values,

$$P(p|p_0) = \frac{1}{2}\int_{-\pi/2}^{\pi/2} \frac{1}{2\pi} \exp\left\{-\frac{p^2 + p_0^2 - 2pp_0 \cos 2(\phi - \phi_0)}{2}\right\} d(\phi - \phi_0) \tag{6.20}$$

$$= \frac{1}{2\pi} \exp\left\{-\frac{p^2 + p_0^2}{2}\right\} \int_{-\pi/2}^{\pi/2} \exp\left\{\frac{2pp_0 \cos 2(\phi - \phi_0)}{2}\right\} d(\phi - \phi_0) \tag{6.21}$$

$$= p \exp\left\{-\frac{p^2 + p_0^2}{2}\right\} I_0(pp_0). \tag{6.22}$$

Here, I_0 is the zeroth-order modified Bessel function, and the $\frac{1}{2}$ factor in the first expression accounts for the 180° rotational symmetry of polarization. This expression is the Rice distribution, which we have derived by assuming Gaussian statistics in q and u, and equal uncertainties on each. To compare with the literature (Serkowski 1958; Simmons & Stewart 1985; Vaillancourt 2006; Quinn 2012), it is useful to also provide this expression with the uncertainties explicitly given,

$$P\left(\frac{p}{\sigma} \,\middle|\, \frac{p_0}{\sigma}\right) = \frac{p}{\sigma} \exp\left\{-\frac{p^2 + p_0^2}{2\sigma^2}\right\} I_0\left(\frac{pp_0}{\sigma^2}\right). \tag{6.23}$$

To find the probability of measuring any true polarization, $P(p)$, one needs to marginalize (integrate the probability distribution) over the true polarization, p_0,

$$P(p) = \int_0^1 P(p|p_0)dp_0 \tag{6.24}$$

$$= \int_0^1 p \exp\left\{-\frac{p^2 + p_0^2}{2}\right\} I_0(pp_0)dp_0 \tag{6.25}$$

$$P(p) = p\sqrt{\frac{\pi}{2}} \exp\left(-\frac{p^2}{4}\right) I_0\left(\frac{p^2}{4}\right). \tag{6.26}$$

Plugging these into Bayes' theorem (Equation (6.14)), and assuming a uniform prior ($P(p_0) = 1$) for the true polarization, we have an analytical expression for the posterior,

$$P(p_0|p) = \frac{P(p|p_0)P(p_0)}{P(p)} \tag{6.27}$$

$$P(p_0|p) = \frac{\exp\left\{-\frac{p^2 + p_0^2}{2}\right\} I_0(pp_0)}{\sqrt{\frac{\pi}{2}} \exp\left(-\frac{p^2}{4}\right) I_0\left(\frac{p^2}{4}\right)}. \tag{6.28}$$

This expression is valid for a single polarization measurement. For multiple measurements, the likelihood (numerator) would need to be modified such that it would be the product of the likelihoods for each individual measurement.

What we derived is an expression for the posterior probability distribution for a measurement of polarization fraction p with an uncertainty σ. The distribution is a function of p_0 such that the peak of the distribution is the most likely value of the true polarization. The shape of the distribution contains information about the uncertainty of the knowledge of the true polarization as well as the most likely value See Figure 6.1.

6.3.2 Estimators

The term "estimator" refers to a function of the measured polarization that approximates the intrinsic polarization of the source, based on some optimization condition. The goal with using an estimator here is to correct for the bias due to the non-Gaussianity of the fractional polarization. As may be expected, there is no single perfect estimator, as the accuracy of the estimator depends on the signal-to-noise ratio. Thus, several estimators are discussed in the literature. We provide a summary discussion here, but the reader is referred to the literature for a deeper exploration of this topic.

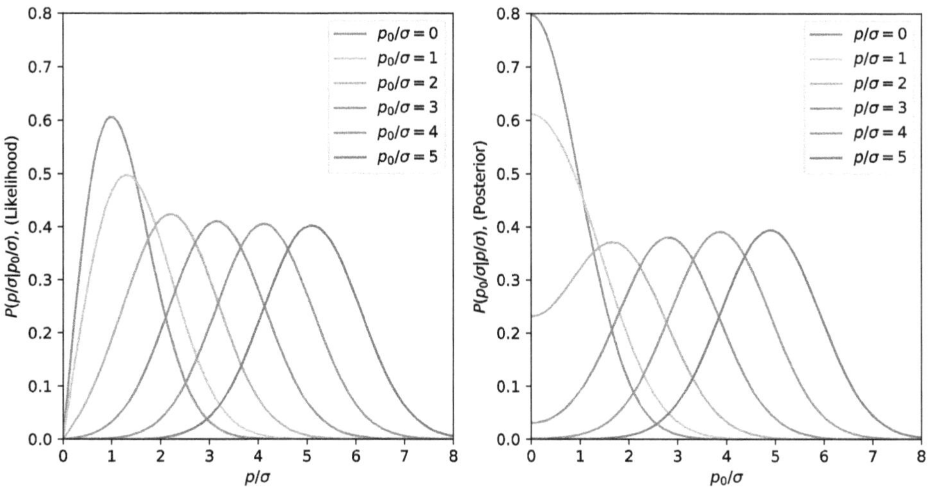

Figure 6.1. (Left) The likelihood as a function of the noise-scaled measured polarization for various values of the noise-scaled true polarization. (Right) The posterior probability density as a function of the true polarization for various values of the measured polarization. Note that the uncertainty (σ) has been explicitly included here to emphasize that these relationships are a function of only the signal-to-noise of the fractional polarization. These figures are reproductions of Figures 1 and 2 from Vaillancourt (2006).

6.3.3 Maximum Likelihood

One potentially useful estimator is to use the *maximum likelihood.* Here we find the value of p_0 that corresponds to the maximum likelihood for a given value of p. To calculate this, one simply needs to set the derivative of Equation (6.23) equal to zero and solve for p_0,[1]

$$0 = \left| \frac{dP(p|p_0)}{dp_0} \right|_{\hat{p} = p_0} \tag{6.29}$$

$$0 = \frac{d}{dp_0}\left[\exp\left\{-\frac{p^2 + p_0^2}{2}\right\}\right] I_0(pp_0) + \exp\left\{-\frac{p^2 + p_0^2}{2}\right\}\left[\frac{d}{dp_0} I_0(pp_0)\right]\Bigg|_{\hat{p} = p_0} \tag{6.30}$$

$$0 = -p_0 \exp\left\{-\frac{p^2 + p_0^2}{2}\right\} I_0(pp_0) + \exp\left\{-\frac{p^2 + p_0^2}{2}\right\}\frac{1}{p_0} I_1(pp_0)\Bigg|_{\hat{p} = p_0} \tag{6.31}$$

$$0 = -\hat{p} I_0(p\hat{p}) + p I_1(p\hat{p}). \tag{6.32}$$

[1] For convenience, we remind the reader that the derivative of the modified Bessel function of the zeroth kind is related to the modified Bessel function of the first kind, $\frac{d}{dx} I_0(x) = I_1(x)$.

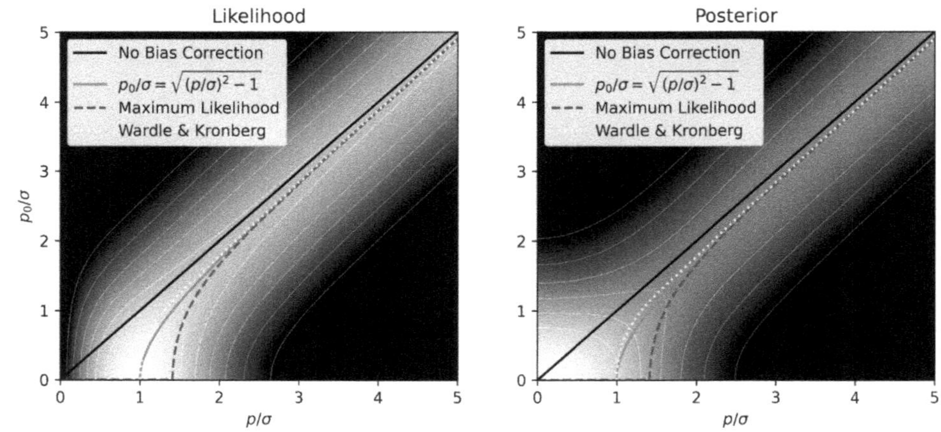

Figure 6.2. The likelihood (left) and posterior probability (right) are shown as functions of both p/σ and p_0/σ. Selected estimators as discussed in the text are shown superposed on the probability distributions. Note also that these are the full distributions from which the curves in Figure 6.1 come.

This result relates the estimator $\hat{p}$ to the measured polarization fraction, p. The maximum likelihood is shown graphically in the left panel of Figure 6.2 as the dashed purple curve. This is a reproduction of Figure 2 from Quinn (2012)—the maximum likelihood is graphically calculated by finding the maximum of each vertical line in the graph.

6.3.4 Wardle and Kronberg Estimator

One can also utilize an estimator for which there is a value of the measured polarization that maximizes the likelihood for a given p_0. This is the estimator utilized by Wardle & Kronberg (1974). In this case,

$$0 = \left| \frac{dP(p|p_0)}{dp} \right|_{\hat{p}\ = p} \tag{6.33}$$

$$0 = (1 - \hat{p}^2)I_0(p\hat{p}) + p_0\hat{p}I_1(p\hat{p}). \tag{6.34}$$

This estimator is shown superposed on the likelihood and posterior functions in Figure 6.2.

6.3.5 Simple Debiasing

Perhaps the most common estimator is the simple analytical expression from Wardle & Kronberg (1974), who suggested that the polarization be corrected by

$$p = \sqrt{p_m^2 - \sigma_p^2}. \tag{6.35}$$

Here, p is the reported polarization, p_m is the measured polarization and σ_p is the estimated uncertainty in the polarization.

6.3.6 Summary of Estimators

As Figure 6.2 shows, none of these estimators are perfect for all usages. Further work has been done. For a complete treatment of the history of estimators, the reader is directed to the references contained in this chapter. The work summarized so far here has assumed a uniform prior on the source polarization. Quinn (2012) has utilized Bayesian decision theory to handle the more general case of arbitrary priors. Montier et al. (2015a, 2015b) have presented detailed analysis of the performance of various estimators. Vidal et al. (2016) has presented an estimator that is of high utility when the angle of polarization is independently constrained. Plaszczynski et al. (2014) has suggested the "modified asymptotic (MAS)" estimator,

$$\hat{p}_{\mathrm{MAS}} = p - \zeta^2 \frac{1 - e^{-p^2/\zeta^2}}{2p}. \tag{6.36}$$

which was used by Planck Collaboration et al. (2020a) for Galactic science. They find a value of the noise-bias parameter of $\zeta^2 < 10^{-3}$ from their data set.

It is quite common to simply debias polarization fractions using the expression in Equation (6.35), which, as can be seen in Figure 6.2, is a good approximation to the Wardle and Kronberg estimator, which Simmons & Stewart (1985) contends to be superior.

One of the drawbacks to simple bias correction of the fractional polarization is that information for low signal-to-noise ratio measurements is lost, because the bias correction assumes these to be zero. For statistical investigations over large data sets, consideration of even low signal-to-noise measurements in the aggregate is sometimes useful.

For example, when investigating polarization fractions as a function of unpolarized intensity, an inverse relationship is often found. There are many astrophysical reasons that may explain this, including the reduction of efficiency of grain alignment in dense cores and a more disordered magnetic field in dense regions due to the magnetohydrodynamic evolution of cloud cores. Debiasing in conjunction with signal-to-noise cuts can lead to its own bias in this relationship, since measurements with both low intensity and polarization are systematically removed in this process. Pattle et al. (2019) have found that in place of debiasing polarization fractions and thresholding, modeling the Ricean distributions can properly assess probabilities of p versus I relationships, by quantifying probable systematic contributions to this relationship.

There are also systematic effects that can produce such a relationship. For example, if one cuts the data from a source at a signal-to-noise threshold, one may selectively eliminate lower polarization measurements in dimmer regions of a cloud for the simple reason that they are noisy. Pattle et al. (2019) has shown the utility of utilizing the full Ricean distribution for estimating the significance of full data sets in reducing bias in such statistical relationships.

6.4 Data Representations and Visualization

Polarization representations are essential as they provide means to both communicate the content of polarimetric measurements and to attain insight into new ideas for physical interpretation. There is no universal method for representing polarization data, and all representations have strengths and weaknesses in that they emphasize certain aspects of a data set and de-emphasize others. In addition, the type of data representation used in dust emission polarimetry of molecular clouds will differ from that used in polarimetry of the cosmic microwave background because of the differences in how the polarization traces the underlying physics. In the case of the former, the primary physics is traced by the polarization orientation at each point in the map. For the latter, the science is directly tied to the spatial frequency content of the polarized component of the signal over the sky. That stated, in most cases, use of multiple representations is helpful in understanding the entirety of the picture.

6.4.1 Maps

Often the data products delivered by polarimeters contain a series of maps corresponding to relevant aspects of the data. One can produce maps of the I, Q, U (and V) Stokes parameters, fractional polarization (p), polarization angle (ϕ; for which it is useful to use a cyclical colormap), polarized intensity ($p \times I$), normalized Stokes parameters, and the uncertainties on any of these quantities. Often these disparate maps provide useful insight, and the utility of a given map is dependent on the science question posed.

6.4.2 "Vectors"

Perhaps one of the oldest and still most common ways of representing a polarization measurement on an astronomical map is by using a line segment that is centered on the measurement. The line segment is typically oriented parallel to that of the measured polarization (electric field). If multiple polarization measurements are placed on a single map, the relative length of the line segment can be used to indicate the fractional polarization. There are various terms informally used for these line segments in the literature that include "vectors", "pseudovectors", and "rods".

Because polarization measurements in the far-infrared through millimeter most often measure *emission* from *magnetically-aligned* dust grains, the inferred magnetic field direction (as projected onto the plane-of-sky) is perpendicular to the polarization orientation. To represent this, polarization vectors are often shown rotated by 90° to represent the magnetic field orientation. In some cases, authors choose to unify the length of the vectors when representing magnetic fields. In other cases, authors choose to use the length of the vectors to represent fractional polarization as in the case of using vectors to represent the polarization orientation. The former practice has the advantage that it reduces the temptation to make inferences about the magnetic field strength across a map; however, it also can give the incorrect impression that the field is more uniform than it is.

Caution should be used in rotating a polarization vectors and interpreting the result as the orientation of the plane-of-sky component of the magnetic field. There is at least one source that exhibits polarization by *absorption* in the far-infrared for the Galactic center cloud Sgr B2 (Dowell 1997). For such sources, differing radiative conditions and magnetic field directions along the line-of-sight require a more detailed comparison with models to understand the magnetic fields.

In addition, at the time of this writing, there is evidence for grains aligned by the so-called k-RAT mechanism, in which dust grains become aligned directly by a radiation field rather than along magnetic fields (Andersson et al. 2022), so some care must be taken before rotating polarization vectors for visualization of an inferred astrophysical magnetic field orientation.

Plotting of discrete vectors is a natural starting point adapted from optical and near-infrared polarimetry. In the early days of the field, single pixel or small array polarimeters made discrete measurements of polarization at specific locations over the sky. With the advent of polarimeters with large detector arrays, both for dust polarimetry and for cosmic background polarimetry, it has become common to create maps of polarization. Pixelization considerations exist in these data sets. For example, a data set may be pixelized significantly below the beam size, such that neighboring pixels in the map are highly correlated. When choosing a vector representation, it is necessary to choose a sampling of the data, since often plotting a vector at each point in a map with a sub-beam pixelization can lead to non-optimal visualizations. Typically, one plots vectors separated by either 1/2 beam (Nyquist-sampled) or a full beam (beam-sampled), depending on the clarity of the choice and other considerations. Since pixelization on small angular scales is typically Cartesian in nature, vectors are typically plotted on a grid. This can lead to an anisotropic representation in which the map looks visually quite different under a rotation of coordinate systems.

6.4.3 Line Integral Contour Maps

With the advent of large filled maps of polarimetry data, the *line integral contour (LIC)* technique (Cabral & Leedom 1993) for visualizing magnetic field geometry has found use. Unlike the plotting of line segments to denote field orientation, which can create bias based on the geometry of the sampling locations, the LIC creates a smooth, aesthetically-pleasing representation of the magnetic field structure that can be semi-transparently overlaid over an image. The LIC is produced via a convolution of white noise with the directional polarization (or inferred magnetic field) information. Figure 6.3 shows the magnetic field data in the Galactic center source, the Sickle at 53 μm as obtained by HAWC+/SOFIA compared to the other visualization strategies described above.

6.4.4 E and B Modes

The Stokes parameters are an extremely useful parameterization comprise an extremely useful parameterization of polarization and due to their Gaussian statistics are nearly universally used when measuring polarization. We have seen

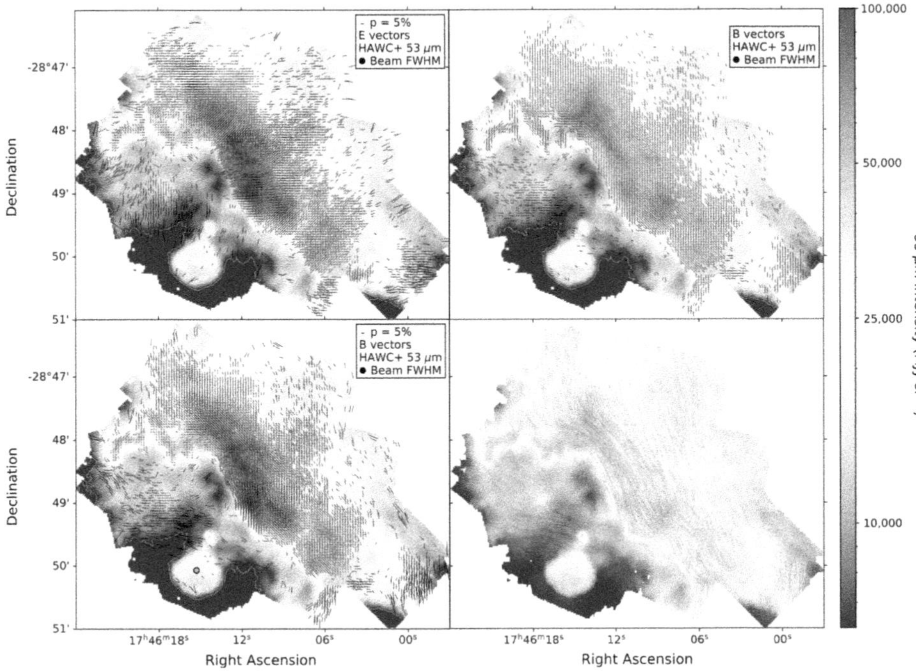

Figure 6.3. Four common ways of displaying polarization information for a discrete source are shown using the "Sickle" molecular cloud in the Galactic center as an example. (Upper left) The polarization is shown for a discrete grid of sampled points (which are typically sampled at a spatial sampling of 1 or 2 points per linear resolution element). The polarization at each point is indicated by a line segment whose direction indicates the direction of the electric field and whose magnitude is proportional to the fractional polarization. (Lower left) Same as for the upper left, but in this case, the direction of the line segments has been rotated by 90° to indicate the direction of the inferred magnetic field. (Upper right) Same as for the lower left, but in this case, all line segments have the same length. (Lower right) Line integral convolution (LIC) representation of the magnetic field direction is shown (see text for details).

that for the case of dust polarimetry, it is useful to parameterize the local polarization in terms of polarization fraction, p, and position angle, ϕ, and the transformations were given in Section 6.2.

In some cases, specifically for the case of cosmology and the polarization of the cosmic microwave background (CMB), a non-local description of the polarization is desired. That is, in contrast to Q and U, which are defined at each resolution point in the sky, the basis set relevant to cosmology are modes of the angular power spectrum over the sky. We will discuss the science driving the desire for such a description in Chapter 7. In short, the physics of the early Universe predict the statistical properties of the polarization.

It is also worth stating that while the Stokes parameters are inherently defined with respect to an orthogonal basis, there is no unique way to define such a basis over the entire sphere of the sky. In the late 1990s, with the WMAP mission on the horizon that would ultimately produce the first all-sky polarization measurements of the CMB, a different mathematical approach to polarization was motivated.

Here we will focus on the mathematical formalism associated with the statistical description of the polarization that is relevant for studies of the CMB. This treatment was first described by Zaldarriaga & Seljak (1997); Kamionkowski et al. (1997). These two papers used different nomenclature for the resulting polarization modes, but here we follow that of the former to be consistent with the common usage of the field.

The polarization, as we have seen, can be described by a 2×2 polarization matrix. For a given direction in the sky described by $\hat{\mathbf{n}}$, a rotation by an angle ψ results in a transformation of the Q and U parameters,

$$\begin{pmatrix} Q' \\ U' \end{pmatrix} = \begin{pmatrix} \cos 2\psi & \sin 2\psi \\ -\sin 2\psi & \cos 2\psi \end{pmatrix} \begin{pmatrix} Q \\ U \end{pmatrix}. \tag{6.37}$$

That is, the quantities Q and U repeat after every 180° rotation of the coordinate system.

This symmetry is captured by creating quantities that are described formally as "spin-2" functions,

$$(Q \pm \mathrm{i}U)'(\hat{\mathbf{n}}) = \mathrm{e}^{\mp 2\mathrm{i}\psi}(Q \pm \mathrm{i}U)(\hat{\mathbf{n}}). \tag{6.38}$$

These quantities can be expanded in spin-weighted bases that span the entirety of the sphere,

$$(Q + \mathrm{i}U)(\hat{\mathbf{n}}) = \sum a_{2,\ell m}\ {}_{2}Y_{\ell m}(\hat{\mathbf{n}}), \tag{6.39}$$

$$(Q - \mathrm{i}U)(\hat{\mathbf{n}}) = \sum a_{-2,\ell m}\ {}_{-2}Y_{\ell m}(\hat{\mathbf{n}}). \tag{6.40}$$

The functions ${}_{\pm 2}Y_{\ell m}$ are the spin-weighted spherical harmonic functions that span the angular space of the sphere for a spin-2 quantity. These are related to the more familiar scalar spherical harmonic functions used to characterize angular momentum. See Figure 6.4 for an illustration of the real parts of these harmonics.

The coefficients of these expansions can be found by integrating,

$$a_{2,\ell m} = \int d\Omega\ {}_{2}Y_{\ell m}(\hat{\mathbf{n}})(Q + \mathrm{i}U)(\hat{\mathbf{n}}), \tag{6.41}$$

$$a_{-2,\ell m} = \int d\Omega\ {}_{-2}Y_{\ell m}(\hat{\mathbf{n}})(Q - \mathrm{i}U)(\hat{\mathbf{n}}). \tag{6.42}$$

Linear combinations of these quantities represent well-defined eigenstates of the parity operator,

$$a_{E,\ell m} = -(a_{2,\ell m} + a_{-2,\ell m})/2 \tag{6.43}$$

$$a_{B,\ell m} = \mathrm{i}(a_{2,\ell m} - a_{-2,\ell m})/2. \tag{6.44}$$

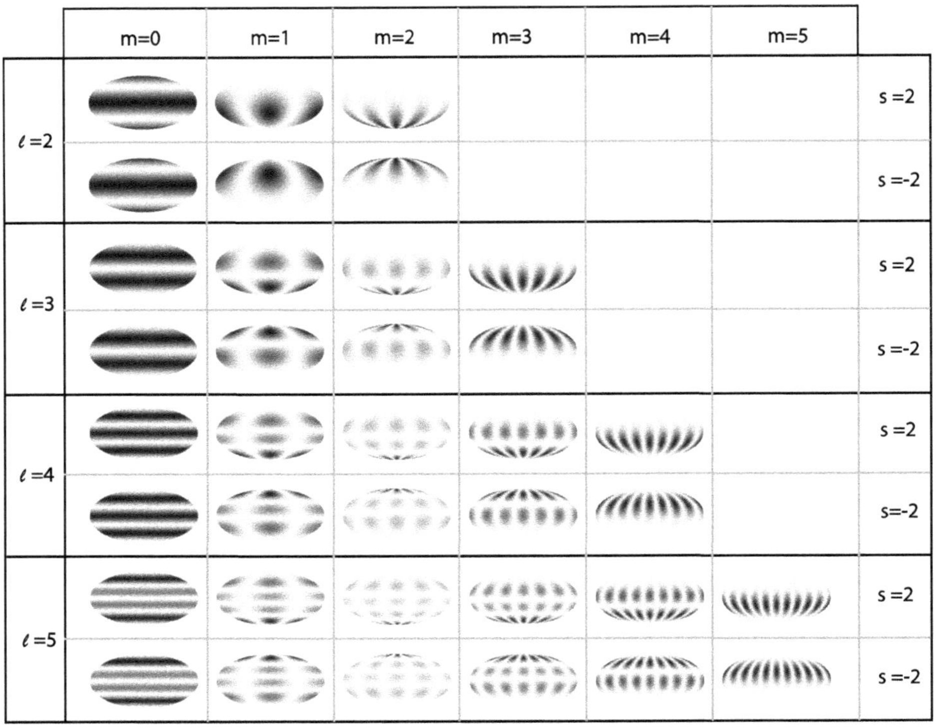

Figure 6.4. The real parts of selected spin-2 spherical harmonics are shown.

The power spectrum of each of these sets of "modes" is found by the autocorrelation of each of the amplitudes,

$$C_\ell^{EE} = \frac{1}{2\ell + 1}\sum_m \left\langle a^*_{E,\,\ell m} a_{E,\ell m} \right\rangle \tag{6.45}$$

$$C_\ell^{BB} = \frac{1}{2\ell + 1}\sum_m \left\langle a^*_{B,\,\ell m} a_{B,\ell m} \right\rangle. \tag{6.46}$$

The fact that two sets of modes are required for the full sky harmonic expansion in polarimetry reflects the spin-2 (pseudovector) nature of polarization. That is, the polarization is invariant under a 180° rotation. The directionality of polarization is such that it requires two distinct sets of modes for a full description. These are the same two degrees of freedom expressed by Stokes Q and U in a local coordinate system. A motivation for the harmonic formalism is that the Stokes parameters cannot be universally defined over the entirety of the sphere. This is articulated in Zaldarriaga & Seljak (1997), "One cannot define a rotationally-invariant orthogonal basis on a sphere." This is mathematically related to the hairy ball theorem theorem from topology, which can be colloquially expressed as, "you can't comb a hairy ball flat without creating a cowlick" (Richeson 2012; Bormashenko & Kazachkov 2017). Additionally, the E and B modes correspond to physics relevant to the CMB.

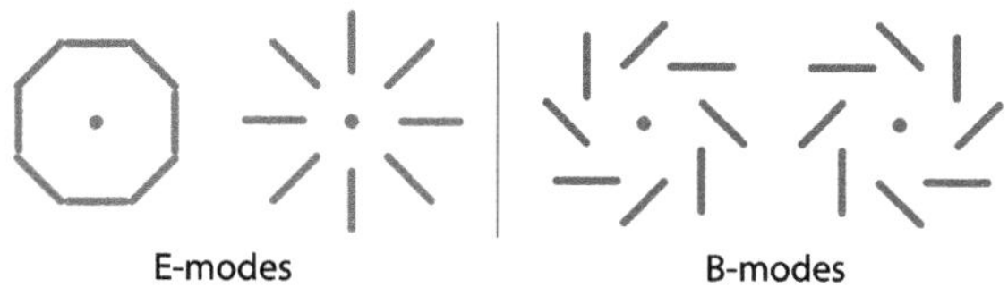

Figure 6.5. An illustration of the symmetries of E- and B-modes at a given point. E-modes are curl-free and have even parity; B-modes are divergence-free and have odd parity.

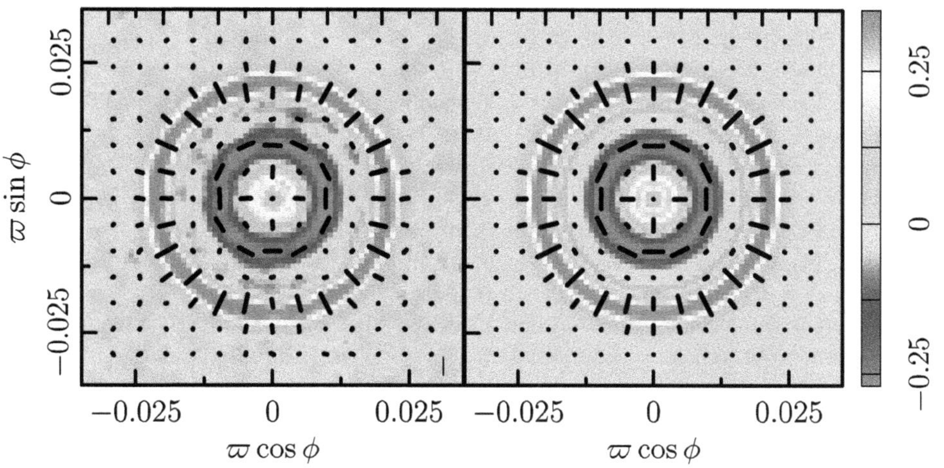

Figure 6.6. (LEFT) Stacked images of intensity and polarization from Planck data are shown for which the E-mode polarization can be seen to be correlated with the temperature anisotropy. (RIGHT) The same image is shown from simulations of the ΛCDM cosmology. The axes scales are in degrees and the colorbar indicating the temperature anisotropy has units of μK. Line segments show the local polarization orientation, and their length is proportional to the fractional polarization (Planck Collaboration et al. 2020b).

Primordial density perturbations can only produce E mode polarization in the CMB; B modes can only be produced by inflationary gravitational waves.

Figure 6.5 shows the symmetries of E and B modes about a point. E-modes have even parity and are curl-free; B-modes have odd parity and are divergence-free. In fact the nomenclature E, B originates as an analogy with electromagnetic fields (Zaldarriaga & Seljak 1997). An alternate G, C notation was introduced (Kamionkowski et al. 1997) for "gradient" and "curl"; however, the former notation has become standard in the field. Figure 6.6 shows a visualization of the E-mode polarization by "stacking" hot spots in the CMB. The E mode polarization resulting from the temperature anisotropy can clearly be seen (Planck Collaboration et al. 2020b).

References

Andersson, B. G., Lopez-Rodriguez, E., Medan, I., et al. 2022, ApJ, 931, 80

Bevington, P. R., & Robinson, K. D. 1969, Data Reduction and Error Analysis for the Physical Sciences (New York: McGraw-Hill)

Bormashenko, E., & Kazachkov, A. 2017, AmJPh, 985, 447

Burgarella, D., Ciesla, L., Sauvage, M., et al. 2024, Proc. SPIE, 13092, 130923B
Cabral, B., & Leedom, L. C. 1993, Proc. of the 20th Annual Conf. on Computer Graphics and Interactive Techniques, SIGGRAPH'93 (New York: Association for Computing Machinery) 263
Dowell, C. D. 1997, ApJ, 487, 237
Emerson, D. T., Klein, U., & Haslam, C. 1979, A&A, 76, 92
Harrington, K., Datta, R., Osumi, K., et al. 2021, ApJ, 922, 212
Hildebrand, R. H., Davidson, J. A., Dotson, J. L., et al. 2000, PASP, 112, 1215
Kamionkowski, M., Kosowsky, A., & Stebbins, A. 1997, PhRvD, 55, 7368
Kovács, A. 2008a, Proc. SPIE, 7020, 70201S
Kovács, A. 2008b, Proc. SPIE, 7020, 702007
Kusaka, A., Essinger-Hileman, T., Appel, J. W., et al. 2014, RScI, 85, 039901
Li, Y., Eimer, J. R., Osumi, K., et al. 2023, ApJ, 956, 77
Mangum, J., Emerson, D., & Greisen, E. 2007, A&A, 474, 679
Miller, N. J., Chuss, D. T., Marriage, T. A., et al. 2016, ApJ, 818, 151
Montier, L., Plaszczynski, S., Levrier, F., et al. 2015a, A&A, 574, A135
Montier, L., Plaszczynski, S., Levrier, F., et al. 2015b, A&A, 574, A136
Pattle, K., Lai, S.-P., Hasegawa, T., et al. 2019, ApJ, 880, 27
Planck CollaborationAghanim, N., Akrami, Y., Alves, M. I. R., et al. 2020a, A&A, 641, A12
Planck CollaborationAghanim, N., Akrami, Y., & Arroja, F. 2020b, A&A, 641, A1
Plaszczynski, S., Montier, L., Levrier, F., & Tristram, M. 2014, MNRAS, 439, 4048
Quinn, J. L. 2012, A&A, 538, A65
Richeson, D. S. 2012, Euler's Gem: The Polyhedron Formula and the Birth of Topology (Princeton, NJ: Princeton Univ. Press)
Serkowski, K. 1958, AcA, 8, 135
Simmons, J. F. L., & Stewart, B. G. 1985, A&A, 142, 100
Sutton, D., Zuntz, J. A., Ferreira, P. G., et al. 2010, MNRAS, 407, 1387
Tegmark, M. 1997, PhRvD, 56, 4514
Tristram, M., Filliard, C., Perdereau, O., et al. 2011, A&A, 534, A88
Vaillancourt, J. E. 2006, PASP, 118, 1340
Vidal, M., Leahy, J. P., & Dickinson, C. 2016, MNRAS, 461, 698
Wardle, J. F. C., & Kronberg, P. P. 1974, ApJ, 194, 249
Wright, E. L., Hinshaw, G., & Bennett, C. L. 1996, ApJL, 458, L53
Zaldarriaga, M., & Seljak, U. 1997, PhRvD, 55, 1830

Chapter 7

Astrophysical Polarization

After examining the instrumentation involved in polarization measurement in astrophysics along with the tools required for reduction and analysis, we turn to the science enabled by the instrumentation. We begin with discussions of selected astrophysical topics that are relevant to polarimetry, including magnetohydrodynamics and dust grain alignment. We discuss several statistical methods for analyzing polarimetry data. Finally, we finish this chapter by summarizing some of the recent techniques and results in astrophysics and cosmology.

7.1 Dust and Polarization in the ISM

Dust grains in the ISM range from sub-nm polycyclic aromatic hydrocarbons (PAHs) to nearly 1 μm.[1] It is generally believed that these grains consist primarily of carbon and silicates (Draine 2011), though there is much debate as to whether the carbon and silicates exist in separate grains or are combined in a single species of dust grain (Hensley & Draine 2023). Generally, aspherical dust grains larger than $\sim 0.05\mu$m (Kim & Martin 1995) can become aligned according to the direction of the ambient magnetic field (or in some cases with the ambient radiation field or pressure gradient). The conditions for alignment vary throughout the ISM, and the details of the grain alignment process are quite an active field of research.

Hall (1949) and Hiltner (1949) first measured polarization of starlight. It was soon surmised (Davis et al. 1951) that the measured polarization was due to anisotropic extinction by dust grains and that these dust grains must be aligned with the direction of the ambient interstellar magnetic fields (Davis 1951).

[1] In a broader context, one can think of dust as ranging from the PAH scale up through the size of rocky planets, but typically, Earth-sized dust does not become aligned by milligauss fields over reasonable timescales, so we'll stick to this definition.

doi:10.1088/2514-3433/ae1022ch7

7.1.1 Grain Alignment Theory

The history of our understanding of grain alignment is rich (Weingartner & Draine 2003; Draine & Weingartner 1997; Lazarian & Hoang 2007a); for an excellent review, see Andersson et al. (2015). For magnetic alignment, the picture seems to be converging with radiative torques from starlight playing a central role in ensuring the suprathermal rotation of the dust grains. This idea, which has become known as "B-RAT" alignment theory[2] was first posited by Dolginov & Mitrofanov (1976) and more recently has been widely studied (Draine & Weingartner 1997). Suprathermal rotation is a key feature for the physics of alignment because it shortens the timescale required for alignment to below the timescale for collisions, which counteract the alignment by re-randomizing the grain orientations. A cartoon representation that captures the main elements of our current understanding of grain alignment is shown in Figure 7.1.

The first element of the alignment process is that radiation (typically in the UV-optical range of the spectrum) is absorbed by the dust grains which typically have a size of $a \sim 0.1\mu$m. The grains are assumed to have a chirality such that they absorb photons having one circular polarization state preferentially over photons having the opposite polarization. When exposed to an anisotropic radiation field, angular momentum is transferred from the radiation field to the dust grains, resulting in the suprathermal rotation described above.

For grains to align, they are required to be aspherical. For any asymmetric body, for a constant angular momentum, the lowest energy spin state is rotation about the axis of largest moment of inertia. To see this, one can consider two rotations ω_1 and ω_2 about principle axes having moments of inertia I_1 and I_2, respectively. Supposing each rotation has an angular momentum L such that

$$L = I_1\omega_1 = I_2\omega_2. \tag{7.1}$$

The energy for each of these states can be written as

$$E_1 = \frac{1}{2}I_1\omega_1^2 = \frac{1}{2}I_1\left(\frac{L}{I_1}\right)^2 = \frac{L^2}{2I_1}, \tag{7.2}$$

$$E_2 = \frac{1}{2}I_2\omega_2^2 = \frac{1}{2}I_2\left(\frac{L}{I_2}\right)^2 = \frac{L^2}{2I_2}. \tag{7.3}$$

If $I_1 > I_2$ then $E_1 < E_2$. Thus, the energetically favored rotation state is that about the moment of greatest moment of inertia. For prolate spheroids (which are the representative grains shown in Figure 7.1), this means the dust grains would tumble "end-over-end, like a well kicked American football," as attributed to Carl Heiles (Hollenbach 1986). Dust may also exist in the form of oblate spheroids, so it may be appropriate to state that the grains exhibit "stable rotation like a well-thrown Frisbee."[3]

[2] "RAT" is short for RAdiative Torques.

[3] From a conversation with B–G Andersson.

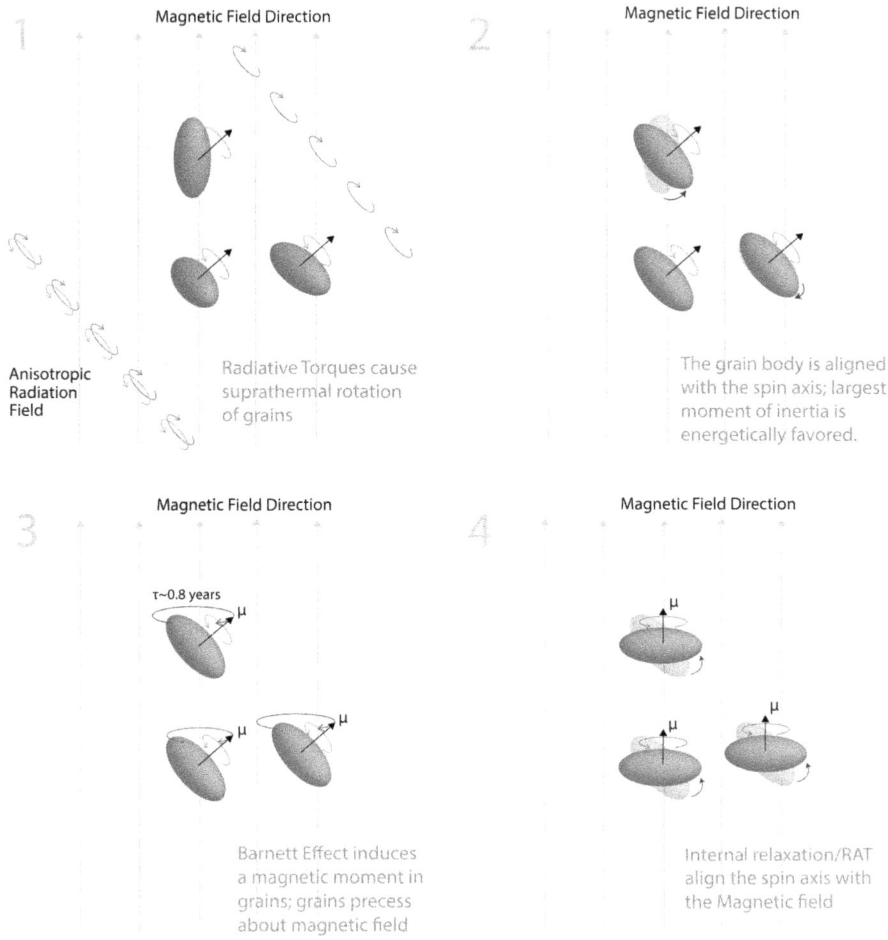

Figure 7.1. Current theory indicates that the partial alignment of dust in the presence of magnetic fields is driven by anisotropic radiation fields imparting angular momentum to chiral dust grains (1). The grains spin preferentially about their axis of greatest moment of inertia (2) and develop a magnetic moment due to the Barnett effect (3). Additional radiative torques and internal relaxation bring the spin axis into alignment with the magnetic field (4).

Once a grain is spinning about its axes of maximum moment of inertia it will manifest a magnetic moment. This is easy to see if the grain has a net charge; however, even neutral grains can become aligned. In fact, the dominant mechanism for inducing a magnetic moment on the grain is the *Barnett effect* (Barnett 1915), in which angular momentum transfer between the macroscopic spin of the grain and the unpaired electrons in the grain result in a partial alignment of these electrons with the spin axis of the grain. Because of this induced magnetic moment, the grain will precess around the magnetic field direction.

Finally, a combination of internal paramagnetic relaxation and additional radiative torques can combine to align the spin axis (and the magnetic moment) of the grain with the magnetic field. The alignment process is illustrated in Figure 7.2.

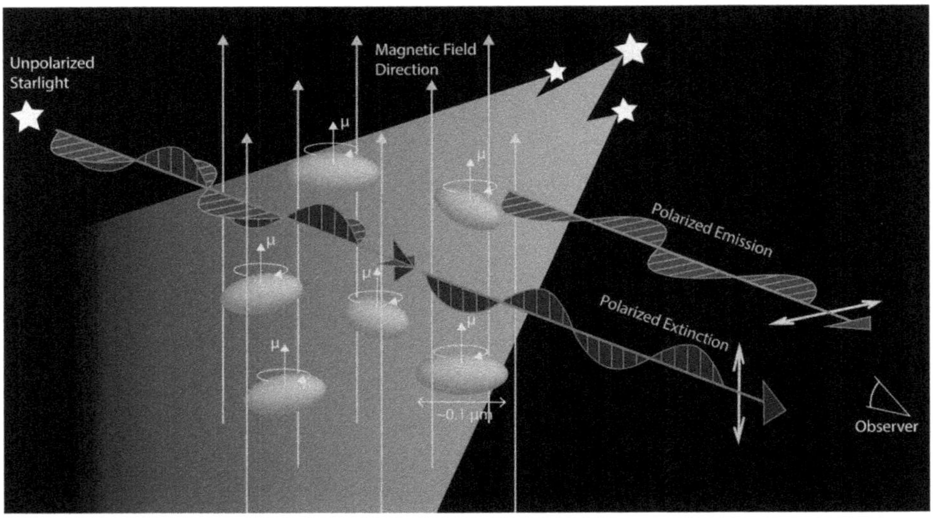

Figure 7.2. Dust grains that are magnetically-aligned are responsible for polarization from the optical through the millimeter parts of the electromagnetic spectrum. In the optical and near-infrared, background starlight undergoes polarized extinction as it passes through magnetized dust. The resulting polarization direction is *parallel* to the magnetic field as projected on the plane-of-sky. At longer wavelengths, the dust grains' temperature is sufficient that polarization by emission is the dominant process. In this case, the measured polarization direction is *perpendicular* to the direction of the magnetic field.

In most of this chapter, we focus on the paradigm of the magnetic field direction as the source of alignment. This is applicable to most observations and interpretations currently; however, we state upfront that grain alignment is an active field of research, and there is some evidence that B-RAT alignment theory is not the only mechanism for grain alignment (or polarization in general). When the radiation field is particularly strong, it is possible for the grains to align according to the radiation direction, $\hat{\mathbf{k}}$ (Lazarian & Hoang 2007a). This mechanism is known as k-RAT alignment. Mechanical alignment is also possible for which the direction of the pressure gradients determine the alignment orientation (Lazarian & Hoang 2007b). Radiative torques have been hypothesized to destroy grains in certain circumstances (Hoang et al. 2019). This process has been dubbed "RAT-D."

Before we end this section, we note an additional polarization mechanism. As we have mentioned before, polarization in astrophysics (or anywhere) involved introducing an anisotropy through which to generate a partial coherence in the photon field. In the above discussion, this has relied on an internal property of the plasma—magnetic field, pressure gradients, radiation anisotropy etc.—to align asymmetric dust grains in a preferred direction. In some cases, the anisotropy can also be introduced in terms of the global geometry with no alignment necessary. This is the case for environments such as protoplanetary and circumstellar disks (Lietzow-Sinjen et al. 2025). In this case, similar to Rayleigh scattering and Fresnel reflection (see Chapter 3), it is the geometry between the source, scatterer and observer that determines the polarization.

7.2 Magnetohydrodynamics in the ISM

7.2.1 Basic Results

The interstellar medium (ISM) is sufficiently conductive such that the dissipation timescale for electrical currents, and consequently for magnetic fields is longer than the timescales over which the combination of turbulent and gravitational forces conspire to form stars. This means that the ISM needs to be considered as a magnetized fluid in modeling star formation. Understanding the role of magnetic fields in star formation—specifically the question of whether magnetic fields are responsible for the low observed rate at which stars are formed—is a primary science driver in studying polarization in the infrared part of the electromagnetic spectrum.

Full treatments of magnetohydrodynamics (MHD) are provided elsewhere (Jackson 1972); however, it is worth briefly reviewing the concept of flux freezing here for completeness, considering its importance to the use of polarimetry for tracing the magnetic field of the ISM. Following Jackson (1972), we use Gaussian units.

Typical treatments consider a neutral, conducting fluid, and the basic formalism considers the electrical conductance, σ, to be uniform in space and time. In this sense, the key equations are Maxwell's equation and the Navier–Stokes equation, with the latter modified to include an electromagnetic force term. In MHD, the key quantities that one solves for are the velocity field, $\mathbf{v}(\mathbf{r}, \mathbf{t})$ and the magnetic field, $\mathbf{B}$. The current density is typically treated as a secondary parameter than can be calculated via Ampere's law; it is more common to treat the magnetic field as the primary parameter in the high-conductivity limit associated with the interstellar medium (Spruit 2013).

The Navier–Stokes equation from fluid dynamics is used in MHD to describe the fluid flow in terms of the density, ρ, pressure, p, and the viscosity of the fluid, η. To incorporate the effect of magnetic fields, it is necessary to include a magnetic force density term, $\mathbf{f}_{\mathrm{EM}}$,

$$\rho\mathbf{g} - \nabla p + \eta\nabla^2\mathbf{v} + \mathbf{f}_{EM} = \rho\frac{\partial\mathbf{v}}{\partial t} + \rho(\mathbf{v}\cdot\nabla)\mathbf{v} = \rho\frac{D\mathbf{v}}{Dt}, \tag{7.4}$$

where $\frac{D\mathbf{v}}{Dt}$ is the convective derivative. The convective derivative, $\frac{D}{Dt} = \frac{\partial}{\partial t} + \mathbf{v}\cdot\nabla$, is also commonly referred to as the Lagrangian or substantial time derivative. In this section, we use the conditions of the ISM to develop the equations that govern the evolution of the magnetic field and derive $\mathbf{f}_{\mathrm{EM}}$ in terms of the magnetic field.

The conductivity of the interstellar medium is sufficiently high that electric fields will be effectively zero for finite current densities (Draine 2011). The relevant equations describing the fields in this scenario are the laws of Faraday and Ampere taken in this limit,

$$\begin{aligned}\nabla\times\mathbf{E} + \frac{1}{c}\frac{\partial\mathbf{B}}{\partial t} &= 0,\\ \nabla\times\mathbf{B} &= \frac{4\pi}{c}\mathbf{J}.\end{aligned}$$

The evolution of the fields can be described by combining Ampere's and Faraday's laws. In the comoving frame of the fluid (referred to as the Lagrangian picture), the current density is proportional to the electric field,

$$\mathbf{J}' = \sigma \mathbf{E}'. \tag{7.5}$$

We wish to solve the equations in the Eulerian picture (lab frame), so we transform the current density and electric field accordingly,

$$\mathbf{J} = \mathbf{J}' + \rho_{\mathrm{e}} \mathbf{v} \tag{7.6}$$

$$\mathbf{E} = \mathbf{E}' - \frac{1}{c} \mathbf{v} \times \mathbf{B}. \tag{7.7}$$

Because the fluid is neutral, the charge density is zero, $\rho_{\mathrm{e}} = 0$. This leads to an expression for the current density as a function of the fields,

$$\mathbf{J} = \sigma \left(\mathbf{E} + \frac{1}{c} \mathbf{v} \times \mathbf{B} \right). \tag{7.8}$$

We next use Ampere's Law and substitute for **J** using the expression in Equation (7.8),

$$\nabla \times \mathbf{B} = \frac{4\pi}{c} \mathbf{J}, \tag{7.9}$$

$$\nabla \times \mathbf{B} = \frac{4\pi}{c} \sigma \left(\mathbf{E} + \frac{1}{c} \mathbf{v} \times \mathbf{B} \right). \tag{7.10}$$

Next, we take the curl of both sides,

$$\begin{aligned}
\nabla \times \nabla \times \mathbf{B} &= \frac{4\pi}{c} \sigma \left(\nabla \times \mathbf{E} + \frac{1}{c} \nabla \times (\mathbf{v} \times \mathbf{B}) \right) \\
\nabla \times \nabla \times \mathbf{B} &= \frac{4\pi}{c} \sigma \left(-\frac{1}{c} \frac{\partial \mathbf{B}}{\partial t} + \frac{1}{c} \nabla \times (\mathbf{v} \times \mathbf{B}) \right) \\
-\frac{c^2}{4\pi\sigma} \nabla^2 \mathbf{B} &= -\frac{\partial \mathbf{B}}{\partial t} + \nabla \times (\mathbf{v} \times \mathbf{B}).
\end{aligned}$$

Rearranging, this leads to a differential equation describing the time variation in the magnetic field as a function of the spatial derivatives of the field,

$$\frac{\partial \mathbf{B}}{\partial t} = \nabla \times (\mathbf{v} \times \mathbf{B}) + \frac{c^2}{4\pi\sigma} \nabla^2 \mathbf{B}.$$

To gain some insight, we first look at the case of a fluid at rest ($\mathbf{v} = 0$). The dynamical equation reduces to

$$\frac{\partial \mathbf{B}}{\partial t} = \frac{c^2}{4\pi\sigma} \nabla^2 \mathbf{B}. \tag{7.11}$$

This is just a diffusion equation for which the field will decay in a time given by

$$\tau = \frac{4\pi\sigma L^2}{c^2}, \tag{7.12}$$

where the quantity L is a characteristic length describing the scale of the system.

In the opposite limit (assuming the conductivity is effectively infinite such that the decay time is long relative to dynamical timescales), one obtains

$$\frac{\partial \mathbf{B}}{\partial t} = \boldsymbol{\nabla} \times (\mathbf{v} \times \mathbf{B}).$$

The curl of the cross product, taking into account that the magnetic field is divergence-free, can be written as (see Appendix B for a derivation)

$$\frac{\partial \mathbf{B}}{\partial t} = (\mathbf{B} \cdot \boldsymbol{\nabla})\mathbf{v} - (\mathbf{v} \cdot \boldsymbol{\nabla})\mathbf{B}. \tag{7.13}$$

Physical intuition about this result can be developed by an argument that parallels that for vorticity in a viscous fluid (Faber 1995). First, we rearrange the above equation to put the negative term on the right-hand side over to the left. We see that the new left-hand side of the equation is simply the convective derivative, $\frac{D}{Dt}$, of the magnetic field,

$$\frac{\partial \mathbf{B}}{\partial t} + (\mathbf{v} \cdot \boldsymbol{\nabla})\mathbf{B} = (\mathbf{B} \cdot \boldsymbol{\nabla})\mathbf{v} \tag{7.14}$$

$$\frac{D\mathbf{B}}{Dt} = (\mathbf{B} \cdot \boldsymbol{\nabla})\mathbf{v}. \tag{7.15}$$

To gain insight, we take the direction of the magnetic field to be the $\hat{\mathbf{z}}$-direction, $\mathbf{B} = B_z\hat{\mathbf{z}}$. This leads to expression for the convective derivatives of the three components of the magnetic field,

$$\frac{DB_x}{Dt} = B_z\frac{\partial v_x}{\partial z} \tag{7.16}$$

$$\frac{DB_y}{Dt} = B_z\frac{\partial v_y}{\partial z} \tag{7.17}$$

$$\frac{DB_z}{Dt} = B_z\frac{\partial v_z}{\partial z}. \tag{7.18}$$

The first two of these equations state that the change in components of the magnetic field perpendicular to a field line are proportional to the gradient of the velocity in that direction. This means that the field lines move with the fluid. The third equation states that the field strength along the direction of the field line increases as the fluid is stretched in this same direction. The lines of magnetic flux are often said to be "frozen" into the medium. That is, the magnetic flux through a Gaussian surface that is moving with the fluid doesn't change.

To characterize a fluid as to which regime is applicable (and to what degree), we define the magnetic Reyolds number,

$$R_{\mathrm{M}} = \frac{v\tau}{L}. \tag{7.19}$$

Here, v is a characteristic velocity of the fluid, L is the characteristic length scale, and τ is the diffusion timescale. Fluids with high magnetic Reynolds numbers are in the flux freezing regime. Those with $R_{\mathrm{M}} \ll 1$ will dissipate their magnetic fields quickly.

We now return to the assumption at the beginning of this section. The electromagnetic force density term $\mathbf{f}_{\mathrm{EM}}$ in the MHD Navier–Stokes equation (Equation (7.4)) can be algebraically manipulated to provide insight as to the nature of the magnetic forces,

$$\mathbf{f}_{\mathrm{EM}} = \frac{1}{c}(\mathbf{J} \times \mathbf{B}) \tag{7.20}$$

$$= \frac{1}{4\pi}(\boldsymbol{\nabla} \times \mathbf{B}) \times \mathbf{B} \tag{7.21}$$

$$= -\frac{1}{4\pi}\left(\frac{1}{2}\boldsymbol{\nabla}B^2 - (\mathbf{B} \cdot \boldsymbol{\nabla})\mathbf{B}\right) \tag{7.22}$$

$$= -\boldsymbol{\nabla}\left(\frac{B^2}{8\pi}\right) + \frac{1}{4\pi}(\mathbf{B} \cdot \boldsymbol{\nabla})\mathbf{B} = -\boldsymbol{\nabla}p_{\mathrm{B}} + \frac{1}{4\pi}(\mathbf{B} \cdot \boldsymbol{\nabla})\mathbf{B}. \tag{7.23}$$

Here, the first term is the gradient of the magnetic pressure, which is related to the square of the magnetic field strength. We have designated this $p_{\mathrm{B}} \equiv B^2/8\pi$. As the fluid is compressed, the field lines are drawn together. This increased magnetic field strength creates a gradient in magnetic pressure that resists further compression. The second term is a magnetic tension, which is a force density along the field lines that tends to resist bending in the field lines.

7.2.2 Sound, Alfvén, and Magnetosonic Waves

For disturbances in the magnetized interstellar medium such as those introduced as a result of turbulence, three distinct types of waves can propagate. For a detailed derivation, the reader is directed to Jackson (1972). Ordinary sound waves can propagate with their wavevectors parallel to the direction of the magnetic field. In this case, the speed is

$$c_{\mathrm{s}} = \sqrt{\frac{dp}{d\rho}}, \tag{7.24}$$

where p is the pressure and ρ is the density. In this case, there is no oscillation in the magnetic fields, since propagation along the field lines does not couple to forces in the MHD equations.

There are also transverse modes in the fluid that propagate along the field direction. In this case, the fluid velocity (and the magnetic field oscillations) perturbations are perpendicular to the magnetic field direction (the wavevector is still parallel to the field direction). These *Alfvén waves* (Alfvén 1942) propagate with a speed,

$$v_A = \left(\frac{B_0^2}{4\pi\rho}\right)^{1/2}. \tag{7.25}$$

These disturbances in the magnetic field have a value related to the velocity of the disturbance in the fluid, **v**,

$$\mathbf{B} = -\frac{k}{\omega}B_0\mathbf{v}, \tag{7.26}$$

which is perpendicular to the undisturbed field direction. Here ω is the angular frequency of the Alfvén waves.

A third type of wave can also present. These are waves where the propagation direction is perpendicular to the undisturbed magnetic field direction. These are *magnetosonic waves* that propagate with a speed,

$$v = \sqrt{v_{\mathrm{A}}^2 + c_{\mathrm{s}}^2}. \tag{7.27}$$

The magnetic field oscillations are parallel to the original field direction,

$$\mathbf{B} = \frac{k}{\omega}v\mathbf{B}_0. \tag{7.28}$$

To close this brief discussion of magnetohydrodynamics, we include a guide to common quantities that appear in the literature. The first nine quantities have been defined in our derivations above. The sonic and Alfvén Mach numbers are the ratios of the fluid speed to the sound and Alfvén speeds, respectively. The Reynolds number is the ratio of the kinetic properties of the fluid to the viscous properties. Generally, turbulence occurs in fluids with $R_{\mathrm{e}} > 1$; whereas if $R_{\mathrm{e}} < 1$, the flow is laminar. Similarly, as seen above, the magnetic Reynolds number, R_{M} is the ratio of the diffusion timescale to the dynamical timescale of interest. When $R_{\mathrm{M}} < 1$, diffusion dominates, and the fluid simply dissipates its magnetic fields resistively. The case relevant to the ISM is $R_{\mathrm{M}} \gg 1$, which corresponds to the flux-freezing condition for which the fields travel with the fluid.

The mass-to-flux ratio, λ provides a means to identify *subcritical* regions in molecular clouds for which magnetic fields dominate the energy density and thus provide support against gravitational collapse ($\lambda < 1$). Conversely, *supercritical* regions of the cloud are those in which gravity dominates over magnetic fields ($\lambda > 1$). Here, we provide the definition of the critical mass-to-flux ratio derived by Crutcher et al. (2004). The critical ratio, $(\frac{M}{\Phi})_{\mathrm{critical}} = \frac{1}{2\pi\sqrt{G}}$ is derived by Nakano & Nakamura (1978).

7.3 Faraday Rotation

In the presence of free electrons, a magnetic field oriented parallel to the propagation direction of radiation can break the symmetry between the effective indices of refraction between right-handed and left-handed circularly polarized light. In such a material, a magnetically-induced resonance beats with the incoming frequency such that for one circularly polarized state, the propagation speed is faster (and the index is smaller) than the mean speed. The other state is slowed relative to this mean speed. This effect is present in the ISM where the free electrons can interact with small (10–100 μG) fields oriented along the line-of-sight to produce significant polarization rotation at radio frequencies. In this case, the relevant resonant frequency is the cyclotron frequency ($\omega_B = \frac{eB}{m_e}$). Faraday rotation can also occur in ferrite materials when a large magnetic field (~kG) is aligned with the propagation direction. In this case, the relevant resonance is the Larmor frequency, ($\omega_L = \frac{eB}{2m_e}$), where we have approximated the gyromagnetic ratio, $g \sim 2$, for the electron (Table 7.1). The latter expression is used in the Faraday polarization rotators described in Chapter 4.

Table 7.1. Commonly-Used Magnetohydrodynamic Quantities and Their Definitions and Relations

Quantity	Description
Pressure	p
Density	ρ
Characteristic Length Scale	L
Fluid Velocity	v
(Dynamic) Viscosity	η
Kinematic Viscosity	$\nu = \eta/\rho$
Fluid Conductivity	σ
Ratio of Heat Capacities	$\gamma = C_p/C_v$
Sound Speed	$c_s = \sqrt{\frac{dp}{d\rho}} = \sqrt{\gamma\frac{p}{\rho}}$
Diffusion Timescale	$\tau \equiv \frac{4\pi\sigma L^2}{c^2}$
Alfvén Speed	$v = \sqrt{B^2/4\pi\rho}$
(Sonic) Mach Number	$\mathcal{M} \equiv \frac{\text{fluid speed}}{\text{sound speed}} = \frac{v}{c_s}$
Alfvén Mach Number	$\mathcal{M}_A \equiv \frac{\text{fluid speed}}{\text{Alfvén speed}} = \frac{v}{v_A}$
Reynolds Number	$R_e \equiv \frac{\text{kinetic terms}}{\text{viscous forces}} = \frac{vL}{\nu}$
Magnetic Reynolds Number	$R_M \equiv \frac{\text{diffusion time}}{\text{dynamical time}} = \frac{v\tau}{L} = \frac{4\pi v L\sigma}{c^2}$
Mass-to-Flux Ratio	$\lambda \equiv \frac{M/\Phi}{(M/\Phi)_{\text{critical}}} = 7.6 \times 10^{-21}\frac{N_{H_2}}{B}$;
Critical Ratio	$(\frac{M}{\Phi})_{\text{critical}} = \frac{1}{2\pi\sqrt{G}}$

Here, we describe the effect of the rotation of linear polarization in radio waves as they travel through the ISM. We follow Rybicki & Lightman (1979) for the general approach. The Lorentz force for an electron is (in SI/MKS units)

$$m_e \dot{\mathbf{v}} = -e\mathbf{E} - e\mathbf{v} \times \mathbf{B}, \tag{7.29}$$

where m_e and e are the mass and charge of an electron.

We will assume that the $\mathbf{B} = B\hat{\mathbf{e}}_3$ is in the z-direction, which represents both the propagation direction and the line-of-sight. The magnetic field in this case is assumed to be constant in time; contributions from the magnetic part of the radiation field will be negligible compared to the electric effect. We can write the circular polarization unit vectors in terms of the Cartesian basis,

$$\hat{\mathbf{e}}_{\pm} \equiv \hat{\mathbf{e}}_1 \pm \mathrm{i}\hat{\mathbf{e}}_2.$$

This allows us to write the electric field in terms of a harmonic quantity with circular polarization,

$$\mathbf{E}(t) = E\mathrm{e}^{-\mathrm{i}\omega t}\hat{\mathbf{e}}_{\pm}. \tag{7.30}$$

Likewise, the velocity also assumes this harmonic form,

$$\mathbf{v}(t) = v\mathrm{e}^{-\mathrm{i}\omega t}\hat{\mathbf{e}}_{\pm}.$$

We note that $\hat{\mathbf{e}}_{\pm} \times \hat{\mathbf{e}}_3 = \pm i\hat{\mathbf{e}}_{\pm}$, and substitute into Equation (7.29),

$$-\mathrm{i}\omega m_e v = -eE \pm \mathrm{i}evB \tag{7.31}$$

$$\omega v = \frac{-\mathrm{i}eE}{m_e} \pm \frac{eB}{m_e}v \tag{7.32}$$

$$\omega v = \frac{-\mathrm{i}eE}{m_e} \pm \omega_B v \tag{7.33}$$

$$v = \frac{-\mathrm{i}e}{m_e(\omega \pm \omega_B)}E. \tag{7.34}$$

Here we have defined the cyclotron frequency, $\omega_B \equiv \frac{eB}{m_e}$. The current density is related to both the velocity and the electric field,

$$J = nev = \sigma E,$$

where n is the density of electrons. Thus, we can write an expression for the conductivity,

$$\sigma = \frac{-\mathrm{i}e^2 n}{m_e(\omega \pm \omega_B)}, \tag{7.35}$$

through comparison with Equation (7.34).

Ampere's law can be rewritten accordingly under the assumption of harmonic fields and low density ($\varepsilon \approx \varepsilon_o$),

$$
\begin{aligned}
\nabla \times \mathbf{H} &= \mathbf{J} + \frac{\partial \mathbf{D}}{\partial t} \\
\nabla \times \mathbf{B} &= \mu_o \mathbf{J} + \mu_o \varepsilon_o \frac{\partial \mathbf{E}}{\partial t} \\
\mathbf{ik} \times \mathbf{B} &= \mu_o \mathbf{J} + \mu_o \varepsilon_o \mathrm{i}\omega \mathbf{E} \\
\mathbf{ik} \times \mathbf{B} &= \mu_o \sigma \mathbf{E} + \mu_o \varepsilon_o \mathrm{i}\omega \mathbf{E} \\
\mathbf{ik} \times \mathbf{B} &= \frac{\mathrm{i}\omega}{c^2}\left(1 + \mathrm{i}\frac{\sigma}{\mathrm{i}\omega\varepsilon_o}\right)\mathbf{E} \\
\mathbf{k} \times \mathbf{B} &= \frac{\omega}{c^2}\left(1 + \frac{\sigma}{\omega\varepsilon_o}\right)\mathbf{E}.
\end{aligned}
$$

This enables the definition of the relative dielectric constant (as discussed in Chapter 3),

$$\varepsilon_r \equiv 1 + \mathrm{i}\frac{\sigma}{\omega\varepsilon_o}.$$

Substituting the definition of σ from Equation (7.35), we can express the dielectric functions for the two circular polarizations as

$$\varepsilon_r^{\pm} = 1 - \frac{\omega_p^2}{\omega(\omega \pm \omega_B)}, \tag{7.36}$$

where $\omega_p^2 = \frac{e^2 n}{m_e \varepsilon_o}$ is the plasma frequency.

At this point, we take advantage of the fact that both ω_p and ω_B are much less than the propagation frequency, ω, to express a limiting form of the dielectric function,

$$\varepsilon_r^{\pm} = 1 - \frac{\omega_p^2}{\omega(\omega \pm \omega_B)} = 1 - \frac{\omega_p^2}{\omega^2\left(1 \pm \frac{\omega_B}{\omega}\right)} \approx 1 - \frac{\omega_p^2}{\omega^2}\left(1 \mp \frac{\omega_B}{\omega}\right). \tag{7.37}$$

The corresponding index of refraction for each circular polarization state is then

$$n_{\pm} = \left[1 - \frac{\omega_p^2}{\omega^2}\left(1 \mp \frac{\omega_B}{\omega}\right)\right]^{1/2} \approx 1 - \frac{\omega_p^2}{2\omega^2}\left(1 \mp \frac{\omega_B}{\omega}\right), \tag{7.38}$$

where we have again used a binomial expansion to first order. We now write the birefringence,

$$\Delta n \equiv n_+ - n_- = 1 - \frac{\omega_p^2}{2\omega^2}\left(1 - \frac{\omega_B}{\omega}\right) - \left[1 - \frac{\omega_p^2}{2\omega^2}\left(1 + \frac{\omega_B}{\omega}\right)\right] = \frac{\omega_p^2 \omega_B}{\omega^3}.$$

The phase delay is then the integral along the line of sight,

$$\delta = \frac{2\pi}{\lambda}\int \Delta n dz, \tag{7.39}$$

$$\delta = \lambda^2 \frac{e^3}{m^2 4\pi^2 \varepsilon_0 c^3} \int n B_z dz. \tag{7.40}$$

Recall from Chapter 2 that the phase delay corresponds to the rotation on the Poincaré sphere, in this case in the *Q–U* plane. The physical rotation is half of this angle,

$$\Delta\theta = \frac{1}{2}\delta = \lambda^2 \frac{e^3}{m^2 8\pi^2 \varepsilon_0 c^3} \int n B_z dz = RM\lambda^2, \tag{7.41}$$

where we have defined the *rotation measure*,

$$RM \equiv \frac{e^3}{m^2 8\pi^2 \varepsilon_0 c^3} \int n B_z dz. \tag{7.42}$$

Given the prevalence of Gaussian units in astrophysics, we can equivalently write the above expression as

$$RM \equiv \frac{e^3}{m^2 2\pi c^4} \int n B_z dz. \tag{7.43}$$

As we can see, Faraday rotation is proportional to λ^2, and it is expressed in units of rad·m^{-2}. Thus observing the same source with multiple wavelengths can allow the determination of the line-of-sight magnetic field direction.

The pre-factor is (Gaussian units) $\sim 2.6 \times 10^{-17} \mathrm{G}^{-1}$. Plugging in representative values allows us to get a sense for the conditions for which Faraday rotation may directly affect observations of polarization. Typical column densities in the diffuse ISM range from $N_{\mathrm{H}} = n\Delta z \sim 10^{19}$–$10^{21}$ cm^{-2} and ISM magnetic field strengths are 10-100 μG. Assuming values in these ranges, the total amount of rotation for a wavelength of 1 cm (30 GHz) can be estimated,

$$\Delta\theta \sim 2.6 \times 10^{-2} \text{ radians}\left(\frac{B}{100\ \mu\mathrm{G}}\right)\left(\frac{N_{\mathrm{H}}}{10^{21}\ \mathrm{cm}^{-2}}\right)\left(\frac{\lambda}{1\ \mathrm{cm}}\right)^2 \tag{7.44}$$

$$\Delta\theta \sim 1.5^\circ\left(\frac{B}{100\ \mu\mathrm{G}}\right)\left(\frac{N_{\mathrm{H}}}{10^{21}\ \mathrm{cm}^{-2}}\right)\left(\frac{\lambda}{1\ \mathrm{cm}}\right)^2. \tag{7.45}$$

So even with the assumption of a strong ISM magnetic field, for 30 GHz and higher, Faraday rotation is negligible. The amount of Faraday rotation as a function of observation wavelength and the mean product of the field along the line of sight and the column density in Figure 7.3.

7.4 Statistical Techniques

7.4.1 Davis–Chandrasekhar–Fermi Method

In independent works in the early 1950s, Davis (1951) and Chandrasekhar & Fermi (1953) proposed that the dispersion of polarization angles measured via starlight

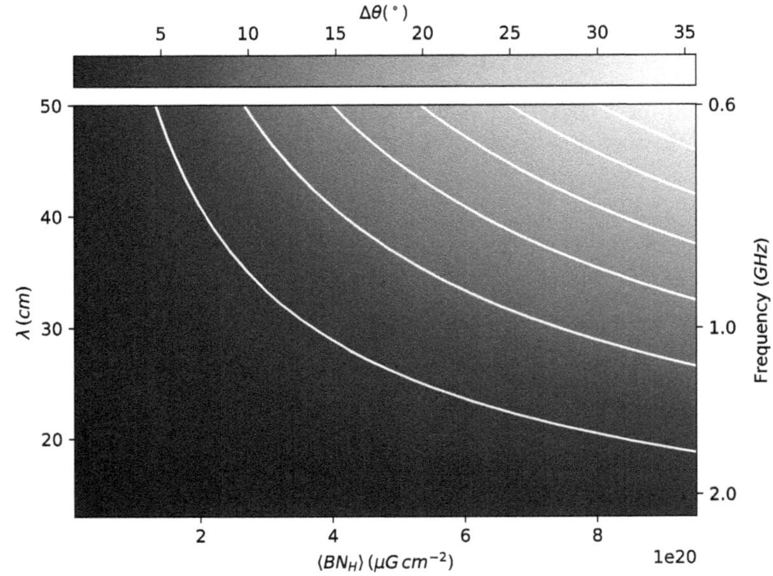

Figure 7.3. The amount of Faraday rotation expected as a function of both wavelength (frequency) and the mean product of the magnetic field and column density is shown.

polarimetry could be used to estimate the strength of the magnetic field in interstellar space. The method, which is commonly known as the Davis–Chandrasekhar–Fermi (DCF) method, relies on the observation that the fields are frozen into the ISM as discussed above. If the fields are strong relative to the turbulent energy in the ISM, they will remain relatively straight. This situation is evidenced by a low dispersion in the measured polarization angles. If, on the other hand, the field is relatively weak, it will be easily pushed around by the turbulent gas motions, and a high dispersion will be measured in the polarization angles. Figure 7.4 shows the basic idea– if the magnetic field is strong, the random motions of the particles do not significantly increase the dispersion of its orientation, as shown on the left panel.

The derivation here closely follows the logic of Chandrasekhar & Fermi (1953), but similar derivations appear throughout the literature. We start with the notion that turbulence in the interstellar medium can be characterized by a velocity dispersion, σ_v^2. This turbulence excites Alfvén wave modes such that the displacement perpendicular to the field direction is

$$y = y_0 \cos k(x - v_A t),$$

where

$$v_A = \frac{B}{\sqrt{4\pi\rho}}$$

is the Alfvén velocity. Here B is the magnetic field strength and ρ is the density. Taking the time and space derivatives of y yield

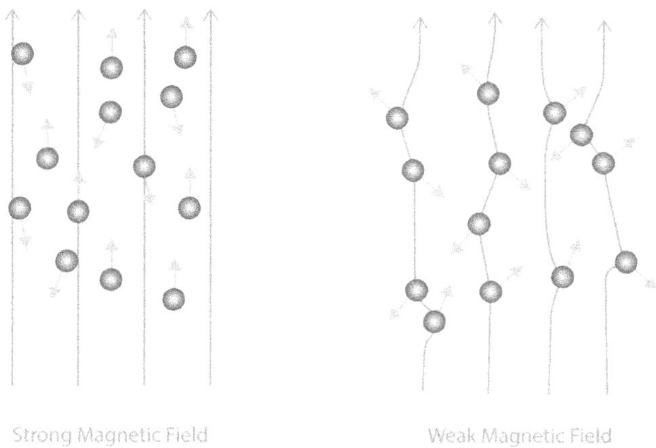

Figure 7.4. The basic concept behind the Davis–Chandrasekhar–Fermi Technique relies on the relative importance of the energy in turbulence versus that in the magnetic field. (Left) In regions of a strong magnetic field/weak turbulence, the magnetic field maintains its structure; whereas in the turbulence-dominated case (right), the magnetic field is dragged by the turbulence, leading to an increase in polarization angle dispersion.

$$\frac{dy}{dx} = y_0 k \sin k(x - v_A t),$$
$$\frac{dy}{dt} = -y_0 k v_A \sin k(x - v_A t).$$

Dividing these equations and then rearranging gives

$$\frac{dy}{dx} = -\frac{1}{v_A}\frac{dy}{dt}.$$

Taking the ensemble variance of each side of the equation gives,

$$\langle \Delta \tan \phi \rangle^2 = \frac{1}{v_A^2}\sigma_v^2.$$

Here, $\langle \Delta \tan \phi \rangle^2$ is the variance in the tangent of the angle of the fields. For small angles, $\Delta \tan \phi \approx \Delta\phi$ and we get the following estimate of the field strength,

$$\langle B \rangle^2 = 4\pi\rho \frac{\sigma_v^2}{(\Delta\phi)^2}, \tag{7.46}$$

where $(\Delta\phi)^2$ is the dispersion of the polarization angle. The turbulence is characterized by the velocity dispersion, σ_v^2, which is obtained from a relevant molecular tracer.

Because the quantities that are available are the line-of-sight velocity dispersion, and the dispersion of the plane-of-sky-projected magnetic field direction, this technique assumes that the turbulence is isotropic. In addition, it is assumed that the dispersion in the polarization traces the variations in the field that are induced by magnetohydrodynamic turbulence. Variations in the field geometry can serve to overestimate the dispersion and hence underestimate the magnetic field strength estimate. To this end, the two-point "structure function" (or "dispersion function, DF") has been developed to fit out first order effects from the large scale field, and to take into account the angular turbulence correlation length, δ, and angular resolution, W, of the observation (Hildebrand et al. 2009; Houde et al. 2011, 2016),

$$1 - \langle \cos [\Delta\phi(\ell)] \rangle = \frac{1}{1 + \mathcal{N}\left[\frac{\langle B_t^2 \rangle}{\langle B_0^2 \rangle}\right]^{-1}} \left\{1 - \exp\left(-\frac{\ell^2}{2(\delta^2 + 2W^2)}\right)\right\} + a_2\ell. \tag{7.47}$$

The values for $\Delta\phi(\ell)$ come from the data and represent the angle difference at a common angular separation, ℓ. Angle brackets indicate averages over the data set. The parameter a_2 estimates the contribution from large scale structure to first order. $\mathcal{N}$ is the total number of turbulent cells in the beam column through the cloud. Houde et al. (2009) relates this column to the thickness of the cloud along the line-of-sight, Δ',

$$\mathcal{N} = \frac{(\delta^2 + 2W^2)}{\sqrt{2\pi}\,\delta^3}\Delta'. \tag{7.48}$$

The parameter $\frac{\langle B_t^2 \rangle}{\langle B_0^2 \rangle}$ is then a proxy for $(\Delta\phi)^2$ in Equation (7.46).

An alternative technique for separation from the turbulent component of the polarization dispersion from variations from the large-scale field geometry is *unsharp (or apodized) masking* (Pattle et al. 2017). In this technique, large scale modes of polarization are removed via a smoothing and subtraction process (high-pass spatial filtering) prior to calculating the dispersion.

Modifications of the classical DCF technique exist. Skalidis & Tassis (2021) propose a modification to the DCF method that takes into account anisotropic turbulence, which they test using simulations. This paper also provides a summary of several variations of the technique as a comparison with their proposed method. Guerra et al. (2023) derives a version of the DCF method for which shear is taken into account and applies it to estimate the magnetic field strength within the central ~5 pc of the Galactic center. They find that without this correction, shear causes an overestimate of the field strength. Myers et al. (2024) provides an additional reference for summarizing details concerning different systematics associated with DCF.

Historically, DCF measurements have relied on limited polarimetric data, thus forcing one to use all of the data for a given source for a single characteristic measurement of the magnetic field. While a good start, conditions, including the polarization dispersion and turbulence can significantly vary over a source, so it is expected that the magnetic field should as well. Guerra et al. (2021) and Hwang et al. (2021) used HAWC+ and Pol-2 data, respectively, to utilize the DCF map across the OMC-1 cloud in order to create a map of the magnetic field strength. This has led to maps of the mass-to-flux ratio across the cloud and thus to initial estimates of the energy balance across the source. Many of these variations in the DCF technique have been coded in the `PolBPy` package,[4] which is publicly available.

7.4.2 Dispersion

The *dispersion* of the polarization refers to the variance of the field (either polarization or magnetic field) direction in the vicinity of a given point in the sky. This is a useful quantity for determining the local variation of the polarization angle, independent of the direction of the field itself. The dispersion is defined at a point $\mathbf{x}$ for a "lag" δ as (Planck Collaboration et al. 2015)

$$\mathcal{S}(\mathbf{x}, \delta) = \left[\frac{1}{N}\sum_{i=1}^{N}(\Delta\phi_{xi})^2\right]^{1/2}, \tag{7.49}$$

where the sum is over N points in an annulus of radius δ and width $\Delta\delta$. $\Delta\phi_{xi}$ is the difference in polarization angles between the center point and point i of the sum,

$$\Delta\phi_{xi} \equiv \phi(\mathbf{x}) - \phi(\mathbf{x} + \boldsymbol{\delta}_i). \tag{7.50}$$

Here, $\boldsymbol{\delta}_i$ is the position of point i in the annulus around point $\mathbf{x}$.

One can recast $\Delta\phi_{xi}$ in terms of the Stokes parameters, using Equation (6.5) and the identity

$$\arctan x \pm \arctan y = \arctan\left(\frac{x \pm y}{1 \mp xy}\right). \tag{7.51}$$

Similarly, one can write

$$\Delta\phi_{xi} = (\phi_x - \phi_i) = \frac{1}{2}\arctan\left(\frac{u_x q_i - q_x u_i}{q_x q_i + u_x u_i}\right), \tag{7.52}$$

as a measure of the variance, $\mathcal{S}$ is a positive-definite quantity like polarization. Therefore, it must also be debiased to avoid systematic overestimate. In this case, since it is itself a type of variance, debiasing also serves to remove in quadrature any contributions from variance due to noisy data,

$$\mathcal{S}^2_{\text{debiased}} = \mathcal{S}^2(\mathbf{x}, \delta) - \sigma^2_{\mathcal{S}}. \tag{7.53}$$

[4] `https://github.com/jorgueagui/polBpy`

The variance in the dispersion can be calculated by direct error propagation using the error in the angle at $\mathbf{x}$, σ_{ϕ_x} and that at each of the other points in the annulus, σ_{ϕ_i},

$$\sigma_{\mathcal{S}}^2 = \frac{1}{N^2\mathcal{S}^2}\left[\left(\sum_{i=1}^{N}\Delta\phi_{xi}\right)^2\sigma_{\phi_x}^2 + \left(\sum_{i=1}^{N}(\Delta\phi_{xi})^2\sigma_{\phi_i}^2\right)\right]. \tag{7.54}$$

7.4.3 Histograms of Relative Orientations and Projected Rayleigh Statistic

To compare with theoretical models for magnetic fluid flow in the interstellar medium, it is useful to compare the magnetic field orientation (as inferred from the polarization data) to the local orientation of the cloud structure.

Such comparisons were first done by Soler et al. (2013) who introduced the *Histograms of Relative Orientation (HRO)*. In the HRO technique, difference angles between the cloud filament direction and the magnetic field are binned. Alternately, the *Projected Rayleigh Statistic (PRS)* essentially quantifies an orientation correlation. Both of these techniques and their associated preprocessing are coded in the `Magnetar` software package.[5]

In addition to the magnetic field information derived from polarimetry, both techniques require an estimate of the gradient of the cloud at each location. The two components of the gradient can be found via the convolution of a Stokes I dust map or column density map with a Gaussian derivative of a specified kernel size. For example, Soler et al. (2013) give an example for a 3×3 Gaussian derivative kernel as

$$\frac{\partial n}{\partial x} = I \circledast \begin{pmatrix} -1 & 0 & 1 \\ -2 & 0 & 2 \\ -1 & 0 & 1 \end{pmatrix} \tag{7.55}$$

$$\frac{\partial n}{\partial y} = I \circledast \begin{pmatrix} -1 & -2 & -1 \\ 0 & 0 & 0 \\ 1 & 2 & 1 \end{pmatrix}, \tag{7.56}$$

where n is the intensity or column density map and the convolution is denoted by $\circledast$.

For both of these techniques, the angle between the magnetic field and the filament orientations must be determined at each position. To do this, one uses the polarization pseudovector, $\mathbf{E}$ and the gradient, ∇n, which are perpendicular to the magnetic field and the cloud elongation, respectively,

$$\phi = \tan^{-1}\left(\frac{|\mathbf{E} \times \nabla n|}{\mathbf{E} \cdot \nabla n}\right). \tag{7.57}$$

For the HRO technique, the values of $\cos\phi$ are binned in a histogram over values between -1 and 1. The area of the histogram between $-0.25 \cos\phi < 0.25$ is denoted $A_{\rm c}$; that for $1 > |\cos\phi| > 0.75$ is denoted $A_{\rm e}$. The degree to which the magnetic field and the cloud shape are aligned is described by the *histogram shape parameter*,

---

[5] `https://github.com/solerjuan/magnetar`

$$\zeta = A_c - A_e. \tag{7.58}$$

Values of $\zeta > 0$ indicate that the field is predominantly parallel to the cloud structures; $\zeta < 0$ indicate a predominantly perpendicular orientation.

An alternate formalism to the HRO, the PRS has the advantage of using all of the information in testing the relative alignment. For the PRS, the *Rayleigh statistic* (Jow et al. 2018) is defined as

$$V' = \frac{\sum_k^N w_k \cos 2\phi_k}{\sqrt{\sum_k^N w_k^2/2}}. \tag{7.59}$$

Here, w_k are the statistical weights for each measurement.

The expression for V' above are sampled over all of the pixels. Since this quantity is unbounded, the pixelization of the image relative to the angular resolution of the system must be taken into account. The oversampling bias can be removed by

$$V = \left(\frac{\theta_{\text{pixel}}}{\theta_{\text{beam}}}\right) V'. \tag{7.60}$$

Values of $V > (<)1$ indicate significant parallel (perpendicular) relative alignment.

An additional statistic for relative alignment is given by Lazarian & Yuen (2018). This *alignment measure*, AM, is defined as

$$\text{AM} = \langle \cos^2\theta - 1 \rangle, \tag{7.61}$$

where the angle brackets indicate an average over the field. Here, θ is the relative orientation of the two vectors fields. This statistic has the advantage of being normalized, such that complete parallel orientation produces a value of 1, and perpendicular orientation yields a value of -1.

7.5 FIR/Submillimeter Polarimetry: Recent Results

We now provide a guide to the current state-of-the-art in far-infrared (FIR) and polarimetric observations by discussing recent advancements in dust emission polarimetry. Ultimately, this topic addresses the multi-scale role of magnetic fields in the life cycles of galaxies. The big picture is as follows. Differential rotation of galaxies enhance seed magnetic fields via a dynamo effect, resulting magnetic field structures that are coherent over kiloparsec scales. Turbulence is injected into the ISM by energetic events such as supernovae and then energy is transferred from this $\sim$50 pc injection scale to smaller scales (McKee & Ostriker 2007). Magnetic fields dynamically affect the flow of matter and are in turn shaped by the dynamical processes, which include turbulence and gravitational collapse. These interactions affect the assembly of molecular structures as clouds gravitationally collapse to clumps and then to the cores where star formation occurs.

How stars form over cosmic time is one of the key physical processes of interest to astrophysicists given the importance of stars in humanity's origin story—stars both provide the mechanisms for producing the raw materials of life from primordial hydrogen and the environments necessary for life to evolve. Thus, the understanding of star formation in this broad context is a high priority. Specifically, one of the puzzles addressed by the study of interstellar magnetic fields is that of the observed low star formation rate within our Milky Way compared to what is expected from gravitational collapse alone.

Our vantage point in the Universe allows for two broad classes of observations relevant to this question. First, observations of external galaxies at various inclination angles can probe the magnetic field holistically across many galaxies at various inclination angles, but with necessarily low resolution. Second, observations of our own Milky Way's ISM provide higher resolution observations of the Galaxy's various environments. Observations of the diffuse interstellar medium that cover large parts of the sky (e.g., Planck and other CMB experiments) at ~arcminute resolution provide information about more uniform sitelines, while pointed observations with sub-arcminute resolution from single-dish telescopes provide information in regions surrounding active star formation. Interferometric observations can provide polarimetric information within the cores of clouds at resolutions of $\sim 1''$.

We begin with a discussion of extragalactic magnetic fields, a subfield that has blossomed in recent years from advances in instrumentation. Then we discuss observations of the diffuse ISM of the Milky Way. We then focus on the magnetic fields in star-forming regions followed by a look at our own Galactic center. We conclude this section with an examination of the wavelength-dependence of polarization, which can provide insights into both grain physics and superposition of different environments along the line-of-sight, both of which are crucial areas of study for a complete understanding of this topic.

The physics of dust and its polarimetric emission is also of interest to the cosmic microwave background (CMB) community for characterization of foregrounds at or above the peak frequency of the CMB. These signatures must be understood and separated from the primordial signal.

As we have seen in Chapter 4, the state-of-the art in this field, as is the case with any subfield of astrophysics, is defined by technology, with discovery driven by advancements in instrumentation. Over the last decade, new instrumentation has blossomed providing observers with unprecedentedly rich data sets. These include balloon-borne instruments such as BLASTpol (Matthews et al. 2014) and PILOT (Misawa et al. 2014) that have provided high-sensitivity polarization data in the far-infrared at ~arcminute resolution, and SOFIA/HAWC+ (Harper et al. 2018) and Pol-2 (Friberg et al. 2016) on the JCMT/SCUBA-2 camera (Holland et al. 2003) that have provided kilopixel-sized arrays with sub-arcminute resolution in the far-infrared and submillimeter, respectively. In addition, though ESA's Planck Surveyor was designed with cosmology as its primary science driver, its data products have provided key large-scale dust polarization information that have worked in concert with higher resolution, pointed experiments to write the next chapter in our understanding of star formation and dust physics.

The magnetic field as inferred from polarization measurements of the dust emission over the full sky by the 353 GHz channel of Planck (Planck Collaboration et al. 2020a) is shown in Figure 7.1 as an LIC overlay on the total intensity (Stokes I) map. These data have been impactful in improving our understanding the magnetic field in the cool neutral phase of the ISM. The diffraction-limited resolution for these observations is $\sim 5'$; however, because the dust intensity varies over the sky, the useful angular resolution depends on the effective beam when coadding over a finite region of the sky. For example, near the Galactic plane, dust emission is sufficiently bright to enable significant signal-to-noise in a single beam $\sim 5'$, whereas near the Galactic poles, many beams must be coadded, resulting in an effective resolution of $\sim 80'$. The Planck data have provided insights across a variety of topics concerning the ISM. The literature describing Planck's intermediate results has been nicely summarized by Planck Collaboration et al. (2020a) in their Appendix A.

7.5.1 Extragalactic Magnetic Fields

Galaxy-scale magnetic fields are understood to occur via dynamo amplification of seed fields; however, there is much to learn about the details of this process. See Borlaff et al. (2021); Lopez-Rodriguez et al. (2022a) for summaries of the current state-of-the-art and set of references to the background literature. In the quest to understand the influence of magnetic fields in the life cycles of galaxies, galaxy-scale magnetic fields provide the large scale framework for the magnetic turbulence that affects dynamics in the interstellar medium for the process of star formation and centers of galaxies. Many of these galactic centers contain active nuclei, and these may be impacted by the magnetic field strength and geometry.

Extragalactic magnetic fields have been studied in the radio and infrared (Packham et al. 1996), tracing the field in the warm, diffuse interstellar medium and dusty sightlines toward stars in the galactic disks, respectively. From these data, we have learned that magnetic fields tend to follow the spiral features of face-on galaxies and are generally parallel to the plane in galaxies that provide an edge-on perspective. Fields are typically coherent over ~kpc scales, and estimates of the strength of the ordered field on galaxy scales, as determined by energy arguments, range from ~few μG in spiral galaxies to $\sim 50\ \mu$G in starburst galaxies. For a review, see Beck (2015).

Radio observations of synchrotron radiation are subject to Faraday rotation, and extinction polarimetry faces the challenge of contamination from Thompson scattering off of free electrons in the host galaxy. As a key complement to these two techniques, far-infrared emission polarimetry traces the magnetic fields in the dense molecular material in the galaxy. At these wavelengths, the material in the galaxy is optically thin, and thus the polarization traces the integrated polarized emission.

Using SHARP, Lee et al. (2015) made the first 350 μm polarization measurement of an active galactic nucleus (AGN), determining the mechanism of the emission to be synchrotron radiation originating from the active region of 3C 279. Since then, the sensitivity in the far-infrared offered by SOFIA/HAWC+ has greatly increased the utility of this technique for studying the field in the dust emission of the cool ISM in nearby galaxies, as evidenced by a significant increase in the number of extragalactic polarization measurements in the FIR and submillimeter. Observations of Cygnus A, a Faranoff–Riley II (FRII) radio galaxy with radio jets, have been made. Ritacco et al. (2017) have measured the polarization of the radio lobes at millimeter wavelengths and determine their emission mechanism to be synchrotron. At far-infrared wavelengths, Lopez-Rodriguez et al. (2018) detected polarized dust emission near the core of the galaxy.

Jones et al. (2019) have mapped the magnetic fields in starburst galaxies M82 and NGC273. M82 shows a magnetic field predominantly perpendicular to the long axis of the galaxy near its center, with a more parallel field toward its edges. The fractional polarization decreases toward the center, suggesting a superposition of a field aligned with the disk in the outer part of the galaxy and a vertical field near the center associated with the starburst. In NGC273, a similar interpretation applies. In this case, the parallel field structure is more pronounced, with only a hint of the vertical field off of the plane, leading to the conclusion that the dust column density associated with any vertical field is much lower than in M82. Lopez-Rodriguez et al. (2021) have utilized a modification of the DCF technique to estimate the strength of the vertical field component associated with the M82 outflow to be 305 ± 15 μG.

Observations at 89 μm of NGC1068, a face-on spiral galaxy, provided evidence of magnetic fields tracing the spiral arms of an external galaxy (Lopez-Rodriguez et al. 2020). Jones et al. (2020) find a similar pattern in the face-on spiral galaxy M51 and a field that is parallel to the plane in the edge-on galaxy NGC891.

A more extensive set of galaxies have been observed by the Survey on extragALactic MagnetiSm with SOFIA (SALSA Legacy program) (Lopez-Rodriguez et al. 2021; Borlaff et al. 2021; Lopez-Rodriguez et al. 2022a; Borlaff et al. 2023; Martin-Alvarez et al. 2024), which measured the magnetic field structures of nearby (<20 Mpc) galaxies at multiple wavelengths. The goals of the SALSA program have been to address the magnetic field geometry as a function of ISM phase, distinguish between theories based on the dynamo mechanism, and trace the fields in outflows and mergers. An image of a selection of the SALSA galaxies is shown in Figure 7.5. As can be seen, the correlation between the magnetic field direction and the spiral structure is strong in face-on cases. Fields also tend to lie within the disk of the galaxies, but the field in central regions of galaxies can be more complex, likely depending on the dynamics of a given galaxy's central engine. SALSA has found a general trend of depolarization in these galaxies. That is, the fractional polarization decreases with increasing column density (Lopez-Rodriguez et al. 2022b). This could be due to field tangling in the dense star-forming regions of the galaxy and is consistent with similar findings in the Milky Way, though the latter case applies to much smaller spatial scales. The combination of these new observations with radio polarimetry are beginning to allow detailed studies of magnetohydrodynamics at the galaxy scale via comparison to models (Martin-Alvarez et al. 2024).

7.5.2 Magnetic Fields in the ISM

One of the long-standing questions in the life cycle of a galaxy is the process by which stars are formed from the interstellar medium. Classically, this involves the

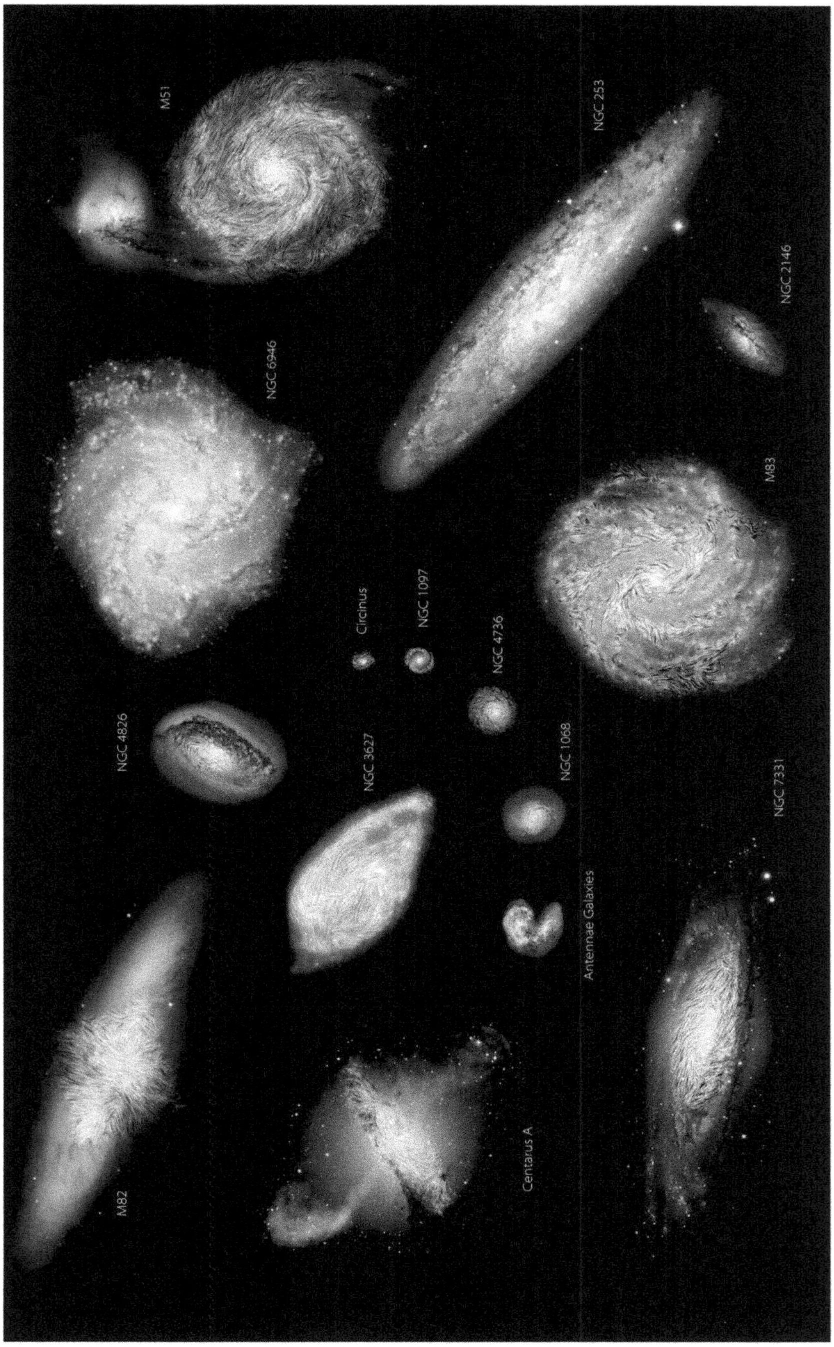

Figure 7.5. A partial collection of galaxies studied in the SALSA project along with the inferred magnetic field direction in each (Borlaff et al. 2023).

gravitational assembly of dense cores from the overdense regions of the diffuse interstellar medium. It is suspected that magnetic fields within the interstellar medium affect this process and, along with turbulence and stellar feedback, are responsible for the observed lower-than-expected rate of star formation within the ISM.

Star formation is understood to occur as a result of magnetized turbulence in the interstellar medium, as energy cascades down from the injection scale, where turbulent modes are excited by energetic events such as supernovae and shocks, down to the dissipation scale (McKee & Ostriker 2007). The theory of magnetized turbulence is complex (Beresnyak & Lazarian 2019), with many of the latest advancement dependent on increased sophistication of MHD simulations.

Polarimetric observations of our own Galaxy can provide insight into the magnetized turbulence at and below the turbulence injection scale. For a majority of lines-of-sight, the dust emission is optically thin, so measurements of polarization are sensitive to the intensity-weighted, plane-of-sky magnetic field integrated along the line of sight. Realizing the goal of fully understanding the configuration and dynamics of the field involves unraveling its three-dimensional spatial configuration along with the associated gas motions. To do this will require the use of polarimetry data in conjunction with other data sets. Optical and near-infrared polarimetry along with data constraining the location along the line-of-sight (e.g., Gaia) will be able to separate regions of different magnetic field directions along the integrated line of sight. Faraday rotation measurements have been used to measure the field along the line of sight (Han et al. 2006), and HI data can be used to assess gas flow directionality as a function of spectral line shift (Clark & Hensley 2019).

As mentioned above, the full-sky map from Planck at 353 GHz (850 μm) has been impactful for addressing the galactic magnetic field (see Figure 7.6). One of the

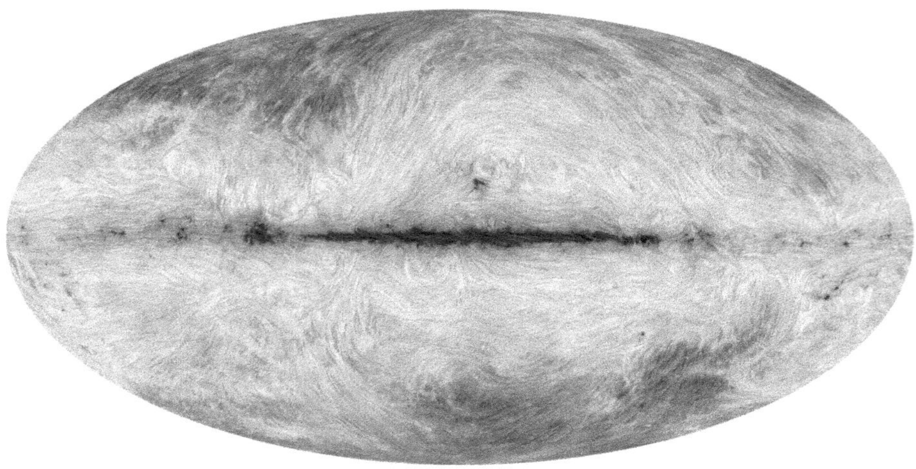

Figure 7.6. The magnetic field as measured by Planck using 353 GHz polarimetry shown as an LIC overlaid on the 353 μm Stokes I map (Planck Collaboration et al. 2020a). Image: ESA and the Planck Collaboration ©ESA.

overall results of the full-sky polarized dust map is that there is approximately twice as much power in E-modes as there is in B-modes (Planck Collaboration et al. 2016a). This has been interpreted to be a result of the Galactic magnetic field's tendency to trace elongated intensity structures in the ISM (Planck Collaboration et al. 2016c). That is, the magnetic field guides the flow of the materials in the diffuse ISM. This E/B asymmetry has been verified in HI emission structures (Clark & Hensley 2019), which has been shown to correlate well with the magnetic field direction inferred from Planck data, but with the advantage of providing information along the velocity axis.

Another insight gained directly from the Planck data is that the maximum fractional polarization at $80'$ resolution is $22.0^{+3.5}_{-1.4}\%$ (Planck Collaboration et al. 2020a). This confirms that dust is generally well-aligned in the diffuse ISM and implies that sightlines with lower fractional polarization, the field is tangled or has a significant component along the line-of-sight.

One of the general trends that is seen in far-infrared and submillimeter polarimetry is that the polarization fraction, p tends to decrease with increasing column density, N_H. This effect was first observed in observations of M17 (Dotson 1996). Possible explanations for such a trend include several factors. First, the magnetic field in denser regions varies over the beam size (often referred to as beam "tangling"), thereby partially canceling in the integrated observation. Second, high column densities could include regions of different magnetic field directions over the line-of-sight, which again could cause cancellation of the polarization from different regions along the sampled volume. Finally, grain and/or environmental properties can differ in denser regions, such that the grains along denser sightlines may be less efficiently aligned. This trend has been observed in other sources as well (Schleuning 1998; Matthews et al. 2001).

In Vela C, Fissel et al. (2010) found that the polarization fraction, in addition to being anticorrelated with column density is also highly anticorrelated with the local dispersion of the polarization angles, $\mathcal{S}$, and suggest that such correlations can be used in conjunction with simulations to address the question of the source of this effect. Planck Collaboration et al. (2016b) find that the anticorrelations between p and both N_H and $\mathcal{S}$ hold over the entirety of the sky. On the scale of star-forming regions, Chuss et al. (2019) find that for OMC-1 in all 4 HAWC+ bands, the stronger of these two anticorrelations was that between p and $\mathcal{S}$. They interpret this as evidence of field tangling, rather than decreasing grain alignment, to be the likely cause of the anticorrelation between p and intensity (I).

One caveat that has been noted is that in noise-limited regions in which there is a signal-to-noise cut on the polarization data, low polarization with low intensity points will be selectively removed and can systematically lead to an anticorrelation between polarization and intensity (which has been used as a common proxy for column density). Pattle et al. (2021) have found that modeling the noise properties and Ricean statistics of non-debiased data leads to significantly more accurate assessment of the p versus I relationship than simple debiasing combined with power law fitting. They test their technique with simulations and apply it to three clumps in ρ-Ophiuchus. In particular, they find that the power law index for ρ-Oph

A is shallower than previously reported. They suggest this is due to better grain alignment in this clump due to its exposure to a higher stellar radiation field.

The Planck data have also led to efforts to test MHD via the statistical relationship between magnetic field orientation and the morphology of cloud structures. This picture of the energy balance of magnetic fields and gravity has been quantified by large-area observations of molecular clouds in the Gould belt. Use of the HRO technique for Planck data (Planck Collaboration et al. 2016b) have revealed that in less dense regions of the ISM ($N_H \leqslant 10^{21}$ cm^{-1}) the field tends to guide the flow of material for the assembly of clouds, as indicated by field that is parallel to the cloud elongation direction. In denser regions, the relative orientation between the field and cloud elongation is predominantly perpendicular as the cloud is able to collapse anisotropically along the field lines (Planck Collaboration et al. 2016b). This has been interpreted to indicate that at low densities, magnetic energy dominates in the assembly of molecular clouds, with a transition point for gravitational dominance. At higher densities, gravitational collapse can occur preferentially for directions parallel to that of the magnetic field. This results in a magnetic field that is preferentially perpendicular to the elongation of the cloud features. High angular resolution polarimetry has confirmed this relative orientation in the ρ-Ophiuchus molecular cloud (Lee et al. 2021; Perry et al. 2025).

Results from SOFIA have indicated that at even higher densities present at the cores of clouds (above $A_v \sim 21$) in the Serpens South molecular cloud, the field returns to an orientation parallel to the cloud structure (Pillai et al. 2020). This indicates a region of high mass-to-flux ratio (supercriticality) for which the magnetic field structure is shaped by gravitational flows.

7.5.3 Star Formation in Molecular Cores

7.5.3.1 OMC-1

One of the most studied molecular cloud systems at these wavelengths, and a favorite early target of each new polarimeter, is the Orion A complex, specifically OMC-1 (M42). Located approximately 390 pc from the Earth (Kounkel et al. 2017), OMC-1 is the closest example of massive star formation and thus enables the most detailed look at this process in terms of both sensitivity and angular resolution. Because of its brightness, it has historically been a common target for polarimeters throughout the development of the field. Indeed, this source was the first observed by most of the early single-pixel instruments (Cudlip et al. 1982; Hildebrand et al. 1984; Dragovan 1986; Clemens et al. 1990; Greaves et al. 1994). These measurements focused on the region surrounding the bright Becklin–Neugebauer (BN) object (Becklin & Neugebauer 1967).

At this distance, $1''$ spans 0.002 pc, so the beam sizes of HAWC+ and Pol-2 are capable of resolving down to the scale of 0.01–0.04 pc, depending on wavelength. Early maps (Schleuning 1998) of OMC-1 have shown an “hour glass” shape to the magnetic field, which has been posited as a qualitative confirmation of magnetically regulated star formation. The basic concept here is that the cloud can more easily collapse along the direction of magnetic field, but in the orthogonal direction, the field is dragged in as it resists the gravitational pull. This results in elongation of the cloud structure that is perpendicular to the original direction of the magnetic field. This is consistent with the statistical findings from the HRO studies described above.

More recent maps of OMC-1 by both SCUBA-2/Pol-2/JCMT (Ward-Thompson et al. 2017) and HAWC+/SOFIA (Chuss et al. 2019) in the submillimeter and far-infrared, respectively, continue to support the magnetically-regulated star formation paradigm. Differences in the hour glass geometry, with the shape being more pronounced in the submillimeter than in the far-infrared. Pattle et al. (2017) use the DCF technique, combined with unsharp masking to remove the large scale fields, to estimate a characteristic magnetic field strength of 6 ± 4.7 mG from the submillimeter polarization dispersion. Chuss et al. (2019) find ~mG values of the field within the region surrounding the BN/KL object using each of 53, 89, 154, and 214 μm far-infrared maps. Their map of the 53 μm inferred magnetic field is the OMC-1 shown in Figure 7.7 as both an LIC overlay (left) and inferred magnetic field pseudovectors (right), each overplotted on the 53 μm Stokes I map.

As mentioned earlier in this chapter, these large data sets have enabled a determination of the magnetic field across the field via the application of the DCF technique at each map pixel by using a subregion surrounding that pixel. Guerra et al. (2021) and Hwang et al. (2021) have done this for far-infrared and submillimeter data, respectively, and find that the dense clumps surrounding BN/KL and the South region are magnetically supercritical ($\lambda > 1$); whereas much of the surrounding region is subcritical. This was a significant step toward quantifying the paradigm of magnetically-regulated star formation in OMC-1.

In addition, the geometry of the magnetic field across OMC-1 corresponds to the morphology of the bipolar explosion discussed by Bally et al. (2011). This outflow is thought to have begun 500 years ago as a result of the orbital decay of a stellar system in the BN/KL region and is notably oriented along the magnetic field direction inferred from polarized emission. Pattle et al. (2017) use energetic arguments to argue that the field is constraining the outflow, rather than the reverse situation in which the outflow is sculpting the magnetic field geometry. Chuss et al. (2019) observed a change of geometry, particularly pronounced in the 53 μm map ($5''$ resolution) in the core of BN/KL region to postulate that at the very center of the explosion, the energy associated with the explosion is sufficient to distort the magnetic field.

7.5.3.2 OMC-2 and OMC-3

The OMC-2 and OMC-3 regions of the Orion A filament that are located just to the north of OMC-1 and have also been mapped in the far-infrared through submillimeter. OMC-3 was first measured at 850 μm (Matthews & Wilson 2000; Matthews et al. 2001) with a measured magnetic field that is aligned with the northern elongated feature of OMC-3 and found consistency with models of a helical field. The magnetic field direction in OMC-3 was verified at 350 μm by Houde et al. (2004) who found that the magnetic field in the bright clumps throughout Orion A varies significantly in direction. Based on magnetic field geometry, Wu et al. (2024) posit that the OMC-2 region is dominated by turbulence due to its twisted field geometry,

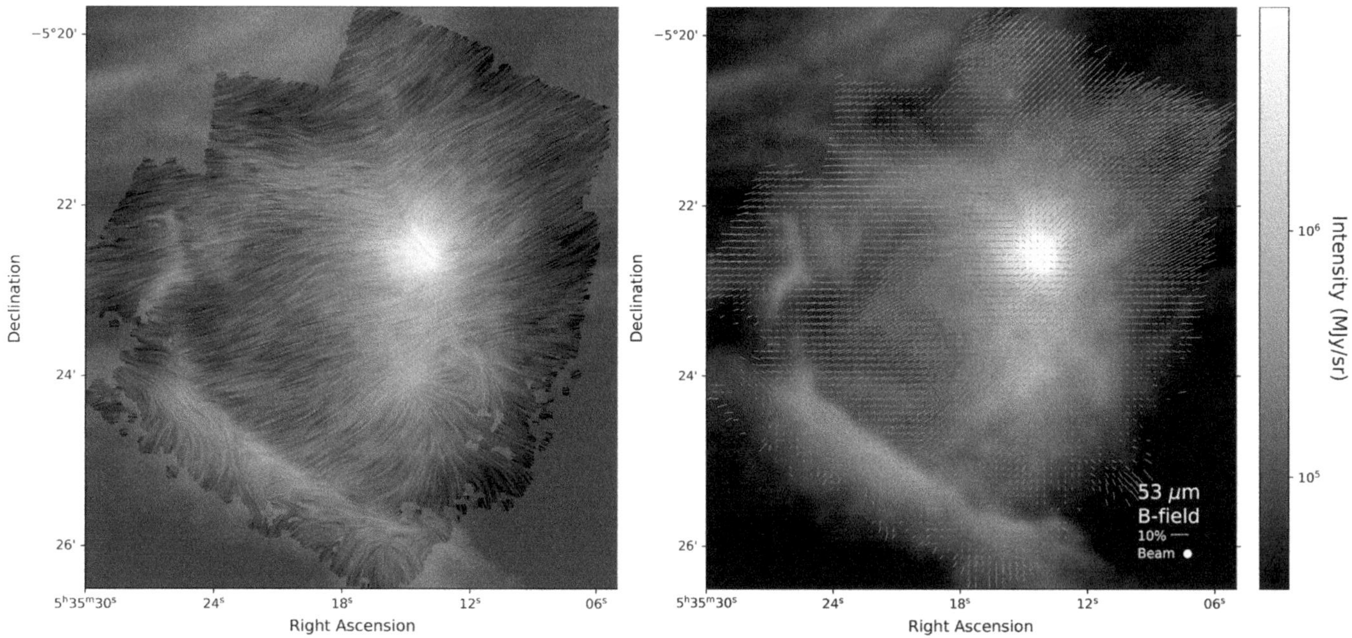

Figure 7.7. Magnetic fields in OMC-1 as inferred by 53 μm polarimetry (Chuss et al. 2019). (Left) The inferred plane-of-sky magnetic field direction is represented by an LIC overlay. (Right) The vector orientation represents the same field geometry information. The magnitude of each vector indicates the fractional polarization. The background image is 53 μm continuum thermal emission from dust as measured by HAWC+/SOFIA.

and OMC-3 is dominated by the magnetic field, as evidenced by the observed straight field lines in this cloud. Using the DCF technique, they find field strengths of 0.45, 0.25, and 0.37 mG for OMC-1, OMC-2, OMC-3, respectively, using 850 μm polarimetry. Zielinski & Wolf (2022) have found magnetic field amplitudes of 202 and 216 μG in OMC-3 using 154 and 214 μm polarimetry.

7.5.3.3 Pillars of Creation

Using SCUBA-2/POL-2, Pattle et al. (2018) measured the magnetic field in the "Pillars of Creation," a star-forming region made famous from the Hubble Space Telescope. They find that the magnetic field direction inferred from 850 μm polarimetry is generally parallel to the filaments (see Figure 7.8), and using the DCF technique, they infer a field strength of 170–320 μG. They conclude that the field geometry was initially shaped by the pillar formation process and is currently slowing the further collapse of the structures.

7.5.4 Galactic Center

The central 150 parsecs is known as the "central molecular zone (CMZ)," due to the presence of $\sim 10^7 M_\odot$ of neutral gas and dust in the region surrounding the $4 \times 10^6 M_\odot$ black hole, known as "Sagittarius A*" (Ghez et al. 2008; Gillessen et al. 2009). This material coexists with hot ionized gas, which is partially manifest as a set of long, narrow synchrotron structures called *nonthermal filaments (NTFs)* (Yusef-Zadeh et al. 1984). These NTFs are oriented predominantly perpendicular to the Galactic plane, are known to trace magnetic fields, and provide an indication of the possible dynamical importance of magnetic field within the Galactic center. The CMZ is coexistent with the $X2$ orbits that are circumscribed by the larger $X1$ orbits that correspond to the Milky Way's bar structure. In fact, the CMZ is framed on either end by the large molecular clouds, Sgr B2 and Sgr C. The former is the most massive molecular cloud within the Milky Way and an early target of polarimetry in the FIR and submillimeter.

This region is of interest due to its uniqueness within the Milky Way. Its elevated temperatures and densities make it distinct from the rest of the ISM, and studying star formation in the region is of particular interest in that it provides a look at this process in an extreme environment. The center of the Milky Way also provides our best opportunity to study the central engines of galaxies at high spatial resolution—an approximately edge-on view of a galactic nucleus at spatial resolutions down to the parsec scale.

Early polarimetry data at 450 μm at $6'$ resolution have revealed that the field in the CMZ is oriented close to parallel to the Galactic plane and interpreted this as evidence for a toroidal geometry (Novak et al. 2003). Sub-arcminute observations found a more complicated field structure at the molecular cloud level (Novak et al. 2000; Chuss et al. 2003). Recently, new observations of the CMZ across these scales have significantly expanded our understanding of the magnetic nature of the dust in the CMZ. On large scales, the ~arcminute resolution of PILOT (Mangilli et al. 2019) and ACTpol (Guan et al. 2021) have confirmed and extended the

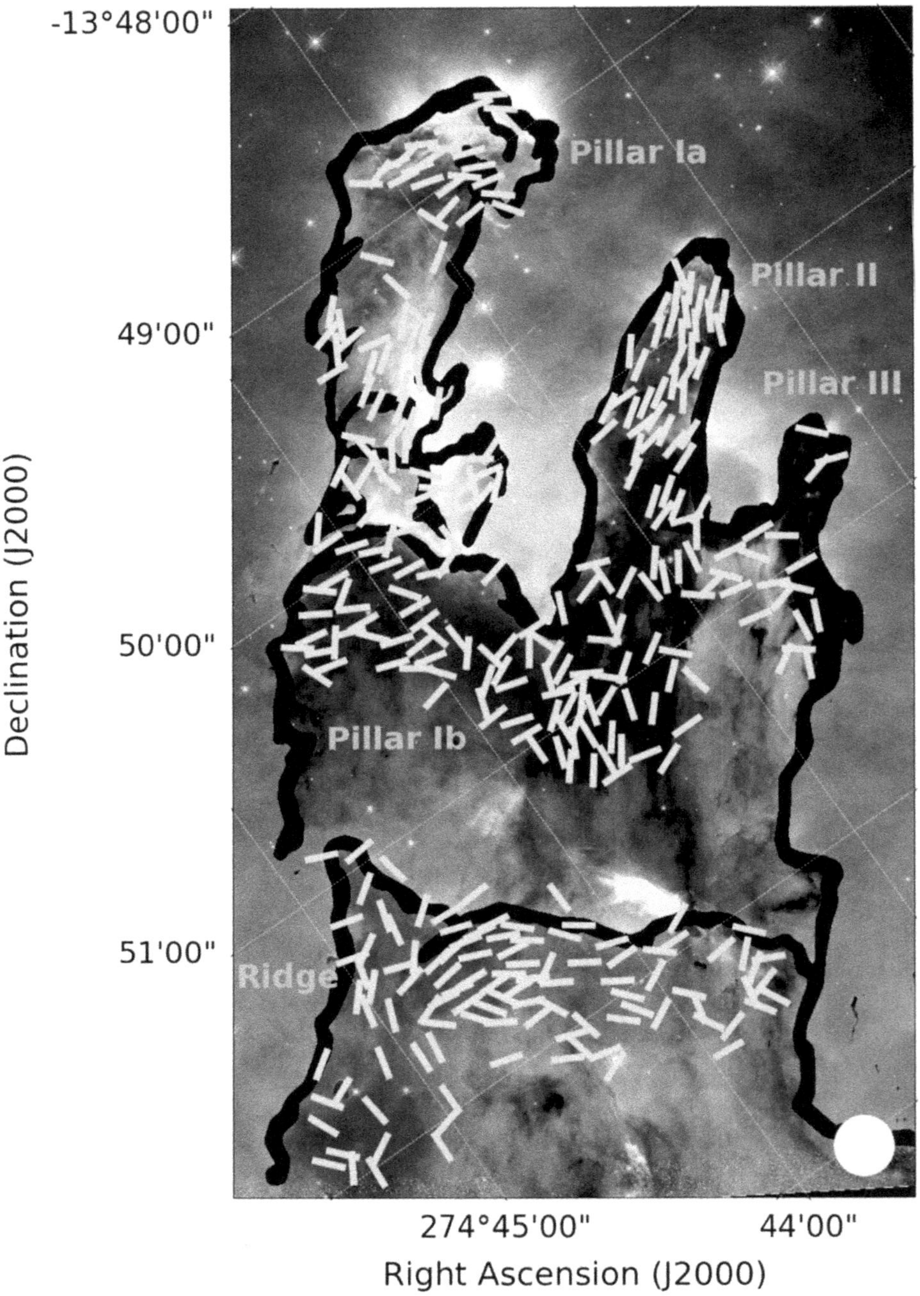

Figure 7.8. Magnetic field orientations are shown superposed over the "Pillars of Creation" (Pattle et al. 2018). The field geometry is shaped by the pillar formation process. The field then supports the columns against radial collapse.

measurements of Novak et al. (2003), which also find agreement with near-infrared absorption polarimetry (Nishiyama et al. 2010). They confirm that the field across the CMZ is fairly uniform with a general orientation of about 20° north of east. This direction is notably nearly perpendicular to the direction of the NTFs. Such experiments are sensitive to large scale modes on the sky, and thus it is likely that much of the polarized emission may come from outside of the central 150 parsecs.

At smaller scales, new surveys from the JCMT/SCUBA-2/Pol2 (BISTRO),[6] (Lu et al. 2024) and SOFIA/HAWC+ (FIREPLACE),[7] (Butterfield et al. 2024; Paré et al. 2024) have uncovered the field in most of the dense cool material in the CMZ at 850 and 214 μm, respectively. The FIREPLACE data are shown in Figure 7.9 as an LIC superposed on a three color image consisting of 70 μm warm dust, 250 μm cool dust (Molinari et al. 2011), and 1.4 GHz radio (Heywood et al. 2022). BISTRO and FIREPLACE achieved similar spatial resolutions—15″ for BISTRO and 20″ for FIREPLACE. Lu et al. (2024) and Paré et al. (2024) have shown that the field on these scales both at 214 and 850 μm is more complex than that detected at arcminute scales, providing support that the field in the CMZ is subject to the cloud dynamics in the central 150 pc.

The field strength has been estimated to be on the order of a few milliGauss, through mostly dynamical arguments (Yusef-Zadeh & Morris 1987a, 1987b; Chuss et al. 2003). Using the DCF technique, Lu et al. (2024) estimates magnetic field strengths in the Galactic center between 0.1 and 1.7 mG.

Butterfield et al. (2024) and Paré et al. (2024) found that the distribution of polarization angles across the CMZ is bimodal, with a component that is roughly parallel to the Galactic plane, consistent with the larger scale ACTpol and PILOT data. The other component is roughly perpendicular to the plane and parallel to the NTFs Heywood et al. (2022) observed at radio frequencies to be representative of a magnetic field perpendicular to the Galactic plane. Paré et al. (2025) have shown that there is also a general trend of the magnetic field following the cloud structure, which they interpret as shearing of the field by differential rotation of the molecular material in the Galactic center potential. Karoly et al. (2025) have found the field geometry to be parallel to multiple streams of matter identified by Kruijssen et al. (2015). These authors find that in areas with star formation activity, the field tends to be perpendicular to the plane, while in areas with a dearth of star formation, the field more closely follows the streams. They point out the potential similarities in the field with polarimetric observations of the starburst galaxies NGC 253 and M82 (Jones et al. 2019).

Such recent results show a picture of the Galactic center magnetic field that is perpendicular to the plane in less dense regions, but dragged along with the rotation in denser regions. Future data analysis combined with modeling efforts are promising for understanding the relation between the fields in the cool dust and warm gas.

---

[6] BISTRO: B-fields in STar-forming Regions Observations.
[7] FIREPLACE: Far-InfraREd Polarimetric Large-Area CMZ Exploration.

Figure 7.9. Magnetic field orientations inferred from the FIREPLACE survey (Paré et al. 2024) are shown as streamlines superposed on a three-color image of the central molecular zone of the Milky Way. The cyan and magenta colors are the 250 and 70 μm maps from Herschel (Molinari et al. 2011) and trace the cool and warm dust components, respectively. The yellow is 1.28 GHz Meerkat data that trace synchrotron emission (Heywood et al. 2022). Several non-thermal filaments, which trace magnetic fields, can be seen in this synchrotron component. (The 3-color background image uses data from European Space Agency (ESA) Herschel Space Observatory and the South African Radio Astronomy Observatory (SARAO) MeerKAT Radio Telescope.)

7.5.5 Wavelength Dependence of Polarization

Up to this point, we have discussed polarimetry results assuming that magnetic fields (B-RAT) is the dominant driver for the alignment mechanism, and thus we can interpret the polarization as tracing the magnetic field geometry. While this is likely a reasonable assumption, the physics of the dust grains, and the alignment process in particular, is not well constrained. The alignment with pressure gradients or radiation direction may dominate under certain conditions in the ISM. Ultimately, to make progress on all of the topics above, it is necessary to improve our understanding of dust physics.

The wavelength dependence of polarization has been used to extract information beyond the field direction and strength. Variations in both the angle and degree of polarization can be useful in probing both dust grain properties and their relevant environments. Typically, such variations indicate either a heterogeneity of grain properties or environmental conditions along the line of sight or a wavelength-dependent change in the physics of the polarization mechanism.

7.5.5.1 Polarization Mechanism

Regions where the polarization angle changes with wavelength can indicate a wavelength-dependent change in polarization mechanism. Dowell (1997) utilized changes in polarization angle and degree as a function of wavelength toward the core of Sagittarius B2 to model the system as a cooler outer cloud with a warmer cloud, each having a distinct magnetic field direction. This work is the first identification of polarization by absorption in the far-infrared, with the inner cloud found to have an optical depth of $\tau_{100\,\mu\mathrm{m}} \simeq 10$. At shorter wavelengths, the polarization mechanism is dominated by polarized absorption of the optically-thick inner source by the aligned grains in the cooler envelope. At longer wavelengths, the polarized emission originates from both the inner, warmer source and the outer, cooler envelope.

Another example of wavelength dependence of polarization angle as an indicator of a transition of the polarization mechanism occurs in the circumstellar disk, DL Tau. HAWC+ measurements (Lietzow-Sinjen et al. 2025) have shown a strong wavelength dependence of the polarization angle, leading to the conclusion that the short wavelengths are magnetically-aligned by the field in the envelope of the source; at longer wavelengths the polarization mechanism is dominated by self-scattering in the disk itself.

7.5.5.2 Testing Dust Physics

Observations along sightlines for which the polarization angle is approximately constant as a function of wavelength provide an opportunity to test the physics of the dust, as these are believed to be less confused by the presence of multiple magnetic field directions superposed along the line-of-sight. The term *polarization spectrum* is typically reserved for the wavelength-dependence of polarization fraction, and it has been extensively utilized as a tool for testing the physics of the dust grains and their environment, including details of the alignment mechanism. In denser molecular clouds, where temperature, density and radiative environments are heterogeneous, the polarization spectrum is highly dependent on the environment and has been used to test grain alignment mechanisms. In translucent and diffuse sightlines, the ISM is much more uniform and the polarization spectrum has been used to test models of the dust composition and shape.

As is the case for the study of magnetic field geometries, early polarization spectrum work focused on observations of molecular clouds. Most of the work done has focused on testing aspects of the RAT paradigm, which posits that grains should attain better alignment in regions that have more exposure to ultraviolet and optical starlight due to either embedded sources or to exposure to the interstellar radiation field (ISRF). Hildebrand et al. (1999) modeled observed falling and rising polarization spectra using (respectively) a superposition of different cloud properties along the line of sight (heterogeneous cloud) and a cloud with multiple grain species (tenuous cloud). Vaillancourt (2002) used far-infrared through millimeter polarimetric observations from a sample of molecular clouds to construct a fiducial polarization spectrum. They observed a minimum in polarization at $\sim$350 μm. Such structure requires grains at multiple temperatures along the line of sight.

Zeng et al. (2013) showed that the shape of the polarization spectrum in M17 varies systematically across a shock front, connecting the nature of the polarization to depend on the environment of the dust. They conclude that the grains that have more exposure to the anisotropic radiation field are better aligned, in accordance with RAT theory.

One of the more peculiar sources studied is that of the environment around the carbon-rich asymptotic giant branch star IRC+10216. The polarization direction has been measured to have a nearly radial direction at 53, 89, and 154 μm (Andersson et al. 2022). These authors suggest that given the strong radiation field, the most likely alignment mechanism is second-order k-RAT, where the grains are externally aligned by the radiation field without alignment between the angular momentum axis and the grain axis (internal alignment). The polarization direction as measured by 850 μm polarimetry differs from the radial pattern seen at shorter wavelengths (Andersson et al. 2024). These authors suggest that this may be due to magnetic alignment at longer wavelengths, and that the grains dominating the polarized signal at longer wavelengths have different composition than those at shorter wavelengths. They suggest more extensive data are required to constrain the physics at work here.

By combining BLASTpol and Planck data, Gandilo et al. (2016) found the polarization spectrum from 250 to 850 μm in the Vela C molecular cloud to be relatively flat, in agreement with the model of Bethell et al. (2007), which considered RAT alignment in a clumpy molecular cloud in the absence of embedded sources. Santos et al. (2019) compared polarization fractions at 89 and 154 μm in ρ-Ophiuchus and found that for denser sightlines there is a relative decrease in polarization fraction between 89 and 154 μm, whereas lower column densities show a rise in polarization fraction between these two wavelengths. The drop in polarization fraction at longer wavelengths is attributed to cooler grains that are less aligned due to shielding from radiation. This is supported by the results of Michail et al. (2021) for OMC-1 in which all four HAWC+ bands (53–214 μm) were used to study trends in the slope of the polarization spectrum. They found a negative slope for cooler regions and positive or zero slopes for warmer ones. The warmer regions selectively contain more well-aligned grains, and the cooler ones contain a superposition of warmer, aligned grains with grains that are less well-aligned due to less exposure to starlight. They suggest that the difference between the trends in OMC-1 and those found in ρ-Ophiuchus may be due to the presence of embedded sources in the latter. Lee et al. (2024) show that simulations of a molecular cloud having similar characteristics to OMC-1 that correspond best to the OMC-1 data (e.g., an overall negative slope for $p(\lambda)$); however, such a model does not reproduce the dependence on the slope of $p(\lambda)$ on temperature. Cox et al. (2025) extends the work of Michail et al. (2021) to other sources measured by SOFIA/HAWC+; however, due to gaps in wavelength coverage, this work uses 3 bands, with OMC-1 as a reference. They find agreement with OMC-1 but mixed results across W3 and M17, indicating more data may be necessary for these sources.

Though these results all generally point to a picture consistent with B-RAT theory, there is clearly more work to do in this area. The complexity of the environs of molecular clouds certainly open the door to systematic uncertainties here. We anticipate that progress in both simulations and in increased fidelity of observations are both crucial for increasing our understanding here. Tram et al. (2021b) test the idea of *Radiative Torque Disruption* (RAT-D) (Hoang et al. 2019) in which large grains can be destroyed by excessive transfer of angular momentum from the radiation field to the grains. These authors find that for multiple wavelengths, the polarization in OMC-1 decreases with increasing dust temperature above 71 K, thus providing evidence to support this concept. A similar result is found in ρ-Ophiuchus (Tram et al. 2021a). Tram et al. (2023) find that the polarization spectrum from 53 to 850 μm for several sightlines in OMC-1 may be more complex than a simple rising or falling spectrum. They have determined that within the context of the RAT paradigm, multiple temperature grains along the line-of-sight must be present to fit the polarization spectra.

Increase in sensitivity has recently enabled study of the polarization spectrum in the diffuse ISM where the polarization spectrum can provide a direct test of dust physics due to the existence of sightlines that are less complicated. Using BLASTpol and Planck data, Ashton et al. (2018) found that the polarization of a translucent cloud is flat, in tension with models from Draine & Fraisse (2009) for which carbonaceous grains are not aligned. Guillet et al. (2018) was able to model these same date for a variety of strengths of the interstellar radiation field and found an improvement in the model fits to the data when aligned carbonaceous grains were included. More recently, Hensley & Draine (2023) have proposed the *astrodust* paradigm, which includes dust grains that include both silicates and carbon. Figure 7.10 shows the modeled polarization spectrum along with measurements from BLASTPol and Planck of the relative polarization.

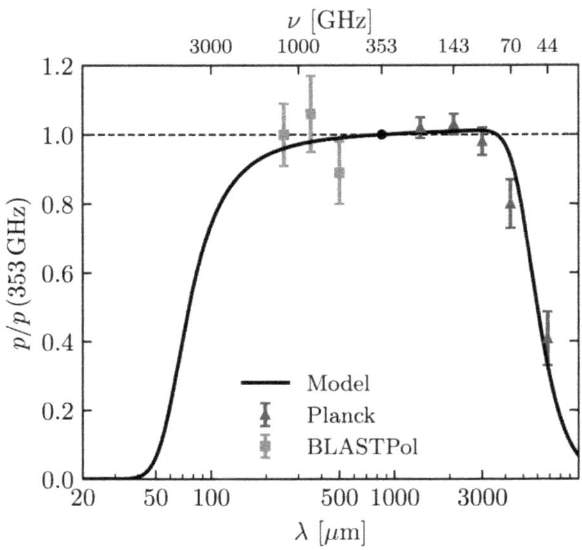

Figure 7.10. The Astrodust + PAH model from Hensley & Draine (2023) is shown superposed on polarization spectrum data from the diffuse ISM from BLASTPol and Planck.

7.6 Polarization of the Cosmic Microwave Background

7.6.1 Introduction: ΛCDM

Thanks in large part to observations of the cosmic microwave background (CMB), the ΛCDM paradigm for an expanding Universe dominated by dark energy (Λ) and cold dark matter (CDM) has emerged as the leading model for the evolution of the Universe. The CMB has provided a wealth of information constraining the six parameters of the model as a result of the community's characterization of the CMB anisotropy in both temperature and polarization. The temperature anisotropy has been measured to near the cosmic variance limit at low multipoles, so polarimetry has become an important frontier in observational cosmology.

There are many excellent references that describe the details of modern cosmology starting with the application of the cosmological principle to general relativity (Dodelson 2003), so we will only summarize the highlights here. In the ΛCDM model, the universe expands and cools from a hot and dense state, undergoing a series of phase transitions during this expansion. The CMB emerges when the universe was ~380,000 years old at the epoch of *decoupling*—the time during which the universe transitions from a plasma to a neutral gas, such that photons no longer appreciably scatter off of free electrons and the universe becomes transparent. As a result, this epoch is often referred to as *the surface of last scattering*. The thermal radiation produced at this epoch becomes redshifted by the expansion and is observed in the microwave part of the spectrum.

The CMB, a nearly perfect blackbody at $T_{\rm CMB} = 2.72548 \pm 0.00057$ K (Fixsen 2009) is uniform over the sky to the level of 1 part in 10^5 (Smoot et al. 1992). The anisotropy in temperature at degree scales and smaller is due to oscillations in the photon-baryon fluid prior to decoupling in the context of a nearly scale-invariant initial dark matter distribution. Larger scale fluctuations represent initial perturbations in this distribution.

The combination of the uniformity of the CMB, the Universe's observed flatness, and the need to explain the origin of the initial dark matter distribution has led to the inflation paradigm (Guth 1981; Linde 1982; Albrecht & Steinhardt 1982). The *inflation* paradigm posits a brief period of exponential expansion early in the Universe's history that addresses these challenges to the standard ΛCDM cosmology (Liddle & Lyth 2000) via a decrease in the Hubble radius.

If inflation is a reality, it would have produced a spectrum of gravitational waves, the amplitude of which is parameterized by r, the ratio of the amplitudes of tensor to scalar modes. This implicitly assumes a flat spectrum for the tensor perturbations. However, it is possible that its spectrum may have a tilt. Because of this, r is typically defined as the ratio of the amplitude of tensor fluctuations to scalar potential fluctuations at $k = 0.002\ \mathrm{Mpc}^{-1}$; however, some papers (Paoletti et al. 2022) consider a potentially tilted spectrum and then specify the scale at which the ratio is defined, r_{k}. The r parameter connects the measurements to fundamental physics in that it is directly proportional to the energy scale of the inflationary potential by

$$V^{1/4} \sim 3 \times 10^{16} r^{1/4}\ \mathrm{GeV}. \tag{7.62}$$

Here, V is the inflationary potential. These gravitational waves leave a faint, but distinct pattern of polarization on the CMB.

7.6.2 Polarization in the CMB

The polarization of the CMB arises from Thomson scattering off free electrons at the time the photons last scatter. Hu & White (1997) present an excellent primer on how polarization is produced in the CMB. Figure 7.11 shows the mechanism schematically. If a free electron lies in a radiation field with a quadrupole anisotropy in a plane perpendicular to the line of sight, it will selectively scatter horizontally-polarized radiation from the up and down directions toward the observer. The lateral directions contribute only vertically-polarized radiation.

The quadrupole anisotropy has multiple origins. It can be produced by the scalar perturbations responsible for most of the TT anisotropy. A quadrupole radiation field can also be produced by inflationary gravitational waves. The radiation field can be polarized by electrons at the surface of last scattering or at the epoch of reionization. For inflationary gravitational waves, the polarization comes from the modes that are entering the horizon at the epoch where the electrons of interest are located.

7.6.3 Measurement of the CMB

The constraining power of CMB anisotropy data is manifest in its statistical properties as a function of scale. The ΛCDM model does not predict the distribution of intensity and polarization on the sky (in map-space). Rather, the model predicts

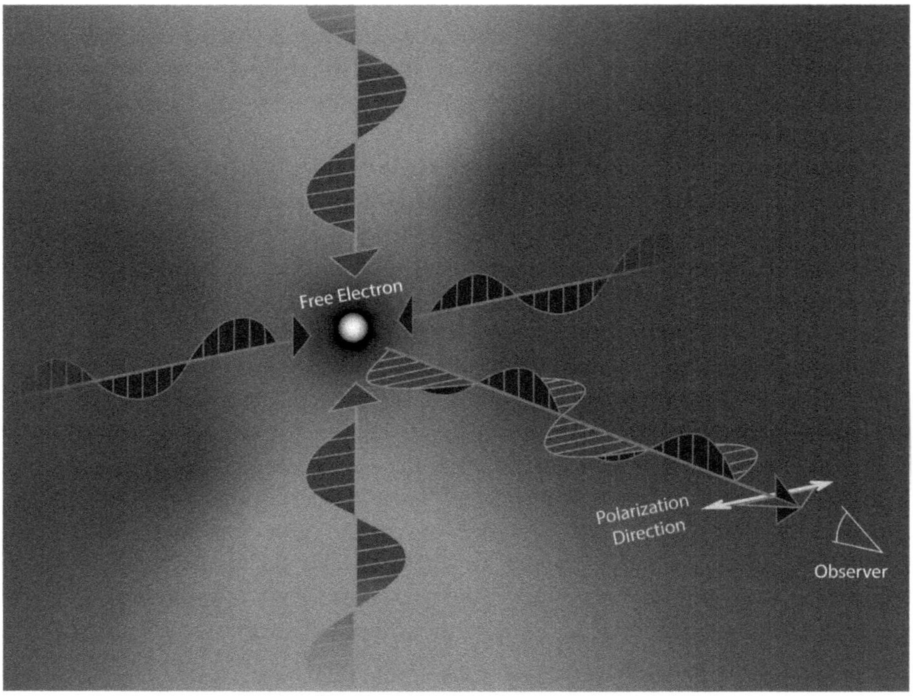

Figure 7.11. The polarization of the cosmic microwave background is caused by Thomson scattering off of free electrons that experience a quadrupole radiation anisotropy at the surface of last scattering. This quadrupole anisotropy can occur as a result of either the primary density perturbations or from primordial gravitational waves produced during an inflationary epoch.

the statistics of the fluctuations as a function of angular scale. This motivates the use of *angular power spectra* as the relevant metric for which to compare models of the early universe with anisotropy data.

The temperature anisotropy (TT) is well-characterized and has provided exquisite determination of the relevant parameters of ΛCDM. The Wilkinson Microwave Probe (WMAP; Bennett et al. 2003, 2013) was the first to characterize the TT spectrum from the quadrupole out to $\ell \sim 1000$, followed by Planck (Planck Collaboration et al. 2020b), which extended the sensitivity of the observations in addition to $\ell \sim 2000$. Ground-based experiments such as ACT (Louis et al. 2025), SPT (Camphuis et al. 2025), and POLARBEAR (Adachi et al. 2022) have extended measurement of the TT spectrum to $\ell \geqslant 3000$. At large scales, the temperature anisotropy determination has reached the cosmic variance limit. Thus, the field has turned to measurements of CMB lensing and polarization at higher sensitivity and resolution.

As seen in Chapter 6, the angular distribution of polarization over the full sky can be characterized by the E and B mode power spectra. This is a valuable basis for describing the polarization that corresponds to the physics involved in producing the anisotropy. The measured spectrum can then be directly compared to those produced by models that capture the physics of the early Universe's expansion and evolution of its constituent components. The primary density (scalar) perturbations produce quadrupole anisotropies that source polarization in the CMB; however, due to the symmetry of these perturbations, only E-modes can be produced as a result. Inflation is predicted to produce gravitational waves that can produce both E and B-modes. Thus, the search for primordial B-mode polarization of the CMB is of high importance to the cosmology and particle physics communities. Discovery and characterization of such as signal would provide both tangible evidence for inflation and a measurement of its energy scale.

The polarization of the CMB is faint. Referenced to the $\sim$2.725 K CMB monopole, the primary (temperature) anisotropy has been measured at $\sim$200 μK (a fractional level of 10^{-5}) The E-mode polarization is $\sim$2 μK (10^{-7}), and then given current measurement limits, the B-mode polarization is $<10 - 60$ nK (10^{-9}).

The low signal level presents multiple challenges. First, the requisite sensitivity is difficult to achieve. The statistical noise is dominated by the unpolarized monopole. This limits the progress one can make by decreasing detector noise, so the technological effort has been spent targeting a combination of large arrays of background-limited detectors and long observing times. Second, the process of measuring the small (polarized) signal in the presence of a larger signal makes systematic control critical. Modulators such as half-wave plates are often employed to convincingly separate the polarization from the unpolarized signal. Beam effects and stray light are additional examples of elements that can produce spurious signals masquerading as primordial polarization.

Figure 7.12 shows the relationship between the temperature and polarization anisotropy. This model was done using the Code for Anisotropies in the Microwave Background (CAMB)[8] software package (Lewis & Challinor 2011). In this illustrative model, we have used parameters from Planck Collaboration et al. (2020c) and have shown a B-mode spectrum corresponding to $r = 0.03$, a value near the current upper limits of measurement constraints (see next section).

The peak in the TT anisotropy curve at $\ell \sim 200$ is the angular size of the sound horizon at the surface of last scattering. Angular scales larger than this are acausal and due to initial conditions; smaller angular scales are characterized by the photon-baryon fluid oscillations during the first 380,000 years. The E and B power spectra show two peaks, one at $\ell \sim 100$, which corresponds to the gravitational wave modes entering the horizon at the epoch of recombination and one at $\ell < 10$, corresponding to the epoch of reionization. The signal from E-modes is significantly larger than that from the as-yet-undiscovered primordially-sourced B modes. However, gravitational potentials between us and the surface of last scattering can lens E modes into B-modes (Zaldarriaga & Seljak 1998). The spectrum of lensing rises from $\ell = 100$ with a peak around $\ell \sim 1000$. These lensing modes have been measured by the BICEP2 Collaboration et al. (2014).

[8] CAMB is based on CMBFAST (Seljak & Zaldarriaga 1996) and freely available: `https:/camb.info`

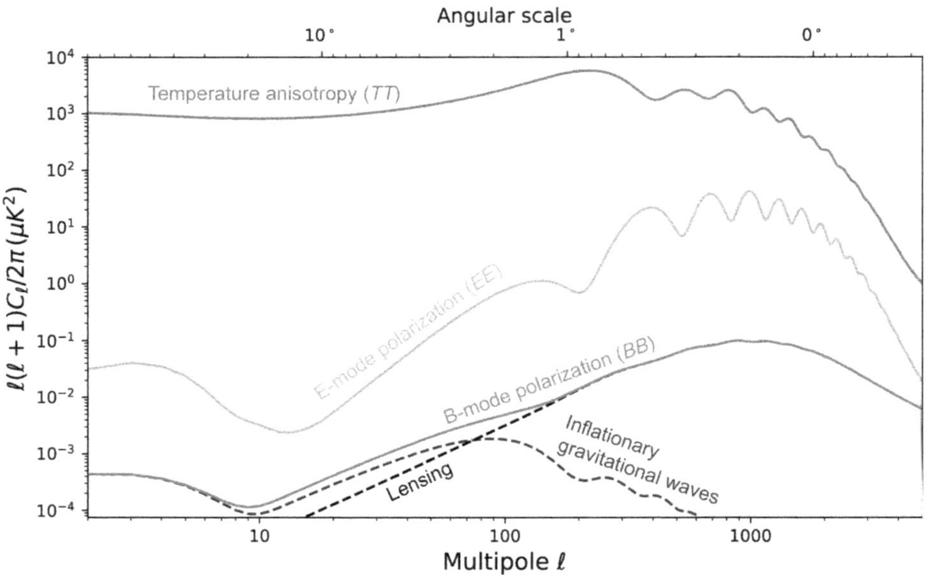

Figure 7.12. An illustration of the temperature and polarization angular power spectra for CMB anisotropies as calculated by the CAMB (Code for Anisotropies in the Microwave Background) software package (Lewis & Challinor 2011). The model uses parameters from Planck Collaboration et al. (2020c). For the B-mode signal from inflationary gravitational waves, the model shows the BB spectrum for $r = 0.03$, near the current 95% confidence limits for multiple experiments.

The E-mode polarization is dominated by the anisotropy produced by scalar perturbations; however, even in the absence of inflationary gravitational waves, it provides additional constraints on the parameters of ΛCDM. In particular, the E-mode spectrum at low ℓ provides a powerful constraint on τ, the optical depth to reionization because it breaks the degeneracy that exists from the TT power spectrum between τ and A_s, the amplitude of the scalar perturbations.

The polarization of the cosmic microwave background (CMB) can provide powerful tests of fundamental physics but can only do so if the polarized emission from our Galaxy can be separated from the primordial signal of the CMB blackbody spectrum. Because of the presence of thermal dust and synchrotron across the entirety of the sky, one of the requirements of a successful test of inflation is the availability of measurements of polarization at many frequencies from ~30–350 GHz, so that the CMB can be isolated based on its spectral energy density. The Planck all-sky CMB polarization map (after removal of foregrounds) is shown in Figure 7.13 (Planck Collaboration et al. 2020b).

The sensitivity gained by large detector arrays and the location of the peak of the CMB (~150 GHz) have motivated the development of CMB observatories operating at 90 GHz and higher, where dust is the main foreground. Thus, understanding the distribution of polarized dust emission across the sky for subsequent removal from the cosmological signal is critical for measuring the polarization of the CMB. The result of this is that from technological and scientific perspectives, efforts to

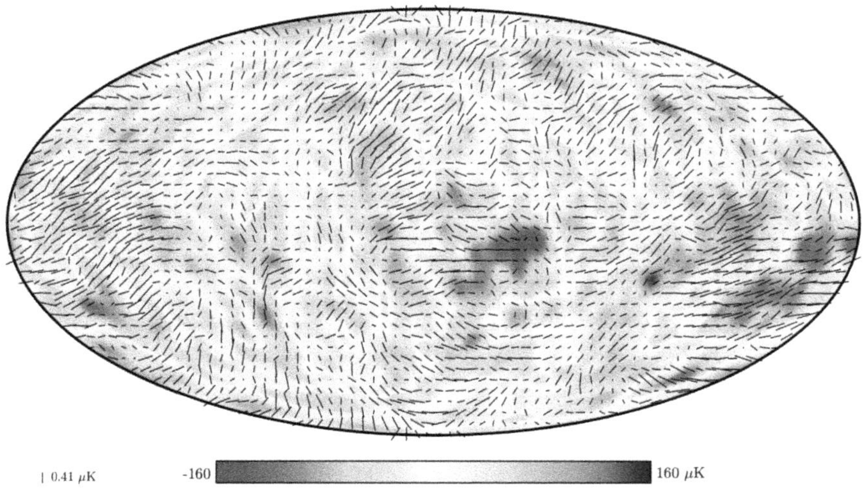

Figure 7.13. Polarization of the cosmic microwave background across the sky, as measured by Planck. Image: ESA and the Planck Collaboration ©ESA (Planck Collaboration et al. 2020b). The background image is the temperature anisotropy and the magnitude of the polarization is shown by the fiducial line segment at the left of the colorbar.

measure the polarization of the cosmic microwave background have had a symbiotic relationship with the interstellar dust and star formation communities. This is manifest, for example, in the tremendous impact the 353 GHz Planck polarization maps have had for dust polarization science, despite the primary science goal of Planck being cosmological.

7.6.4 Observations

7.6.4.1 Testing the Λ CDM Paradigm

The first detection of polarization of the CMB was made by the Degree Angular Scale Interferometer (DASI; Leitch et al. 2002) via measurement of the *TE* power spectrum (Kovac et al. 2002) from the Amundsen-Scott South Pole Station. Measurement of the polarization angular power spectra at scales larger than the sound horizon at recombination probes physics at the reionization epoch. This has led to the E-mode polarization as a key tool for the measurement of the optical depth to reionization, τ. The *Wilkinson* Microwave Anisotropy Probe (*WMAP*) utilized measurement of the *TE* power spectrum to enable the first measurement of τ (Kogut et al. 2003) at large angular scales (low ℓ). Since this initial measurement of $\tau = 0.17 \pm 0.04$, improved E-mode measurements and analysis strategies have refined this estimate. The latest results combine Planck and CLASS data and constrain this value to be $\tau = 0.053^{+0.018}_{-0.019}$ (Li et al. 2025).

The *EE* power spectrum has been measured by WMAP (Page et al. 2007) and later refined by Planck (Planck Collaboration et al. 2020d) and ground-based instruments: South Pole Telescope (SPT; Camphuis et al. 2025), BICEP/Keck (Tristram et al. 2022), POLARBEAR (Adachi et al. 2022), and the Atacama Cosmology Telescope (ACT; Louis et al. 2025). A summary of the measurements of the power spectra are shown in Figure 7.14, which has been compiled by Louis et al.

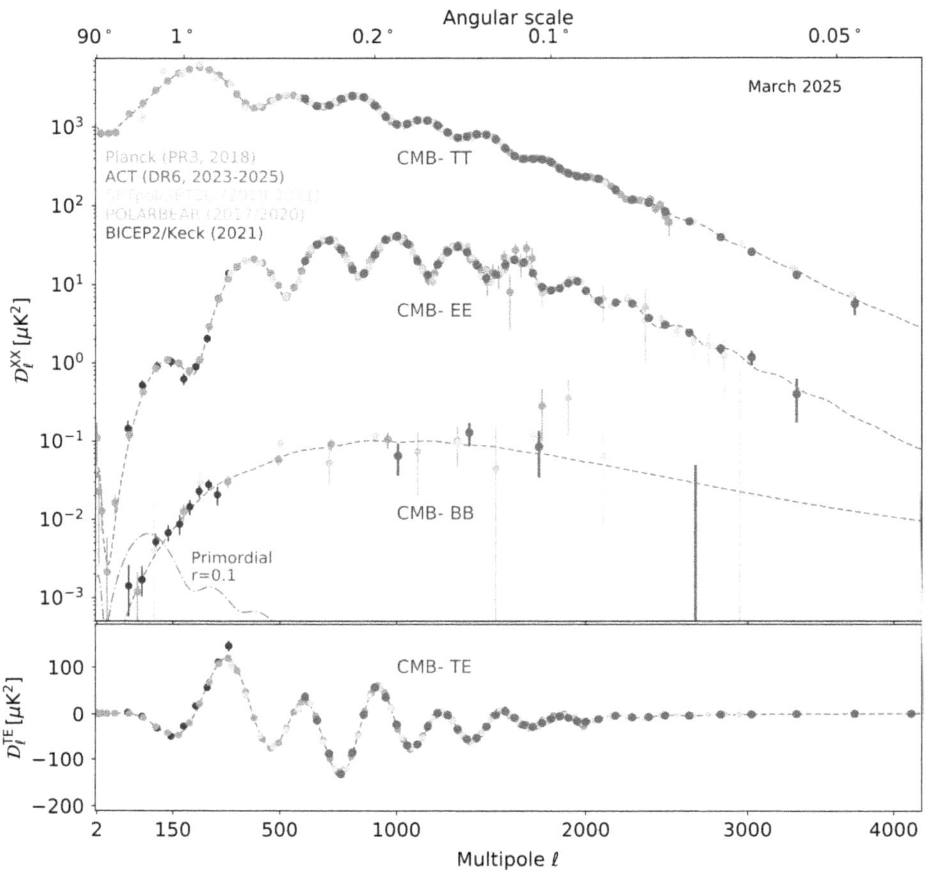

Figure 7.14. A compilation of recent CMB temperature and polarization power spectrum measurements (Louis et al. 2025).

(2025). The *EE* spectrum is well constrained by current data. The *BB* spectrum has revealed a lensing profile that is consistent with the level of the *E*-modes. So far, no primordial *B*-mode component has been detected.

As mentioned above, the amplitude of inflationary gravitational waves is proxied by the tensor-to-scalar ratio, r. Despite some initial excitement in 2014 of a reported detection that later turned out to be foreground dust (BICEP2 Collaboration et al. 2014), no significant detection of $r > 0$ has been made to date. Progress has been steady on the measurement of this parameter. In the early 2000s, CMB temperature provided the most powerful constraints on the polarization (Peiris et al. 2003). As polarization sensitivity increased, polarization began to be the dominant constraint. The current upper limit on the tensor-to-scalar ratio is $r \lesssim 0.032$, which is determined using a combination of Planck, BICEP/Keck, and BAO data (Tristram et al. 2022).

7.6.4.2 Circular Polarization

Though most of the focus on CMB polarization has been on linear polarization measurements given their high value for constraining physics and their measurability, circular polarization has been considered both from early Universe physics (Alexander & McDonough 2019) and through *Faraday conversion*. The latter scenario can occur via the conversion of linear polarization of the CMB into circular via transmission through a birefringent medium. Several papers explore the possible sources of this birefringence via both physics induced by the CMB anisotropy itself (Montero-Camacho & Hirata 2018) and through tensor and vector perturbations (Inomata & Kamionkowski 2019). In all cases, the circular polarization is expected to be much smaller than the linear.

CMB experiments have made progress in measuring the circular polarization. In each case, coupling between the linear and circular polarization states in modulators of the instrument was used to place limits on the power spectrum of V. The Milan Polarimeter (MIPOL; Mainini et al. 2013), operating at 30 GHz, placed limits of V at $\ell \sim 8$–26 of $\sim 10^5$ μK^2 using coupling between Stokes U and V in its correlator. SPIDER (Nagy et al. 2017) used linear-to-circular coupling in its half-wave plate to constrain the upper limit of circular polarization from $\ell \sim 33$–307 to 141–255 μK^2 at 150 GHz. The most stringent upper limit to date is from CLASS, which utilized a VPM with significant sensitivity to Stokes V to provide a null channel for systematic control. After removing circular polarization from the atmosphere at 40 GHz (Petroff et al. 2020), they place limits on C_ℓ^{VV} from $1 < \ell < 120$ of 0.4 to 13.5 μK^2 (Padilla et al. 2020).

7.6.4.6 Cosmic Birefringence

The nature of dark energy remains one of the key puzzles of modern cosmology. Making up more than 75% of our Universe's energy content, this entity is characterized by a negative energy density that is accelerating the expansion of the Universe. One possible origin for dark energy is that it originates from a parity-violating pseudoscalar field. Such a field would cause an electromagnetic wave experience different phase velocities for right- and left-hand circular polarization states. This will create a (circular) *cosmic birefringence* for the CMB that will rotate its polarization at each point in space through an angle β, as seen schematically in Figure 7.15. Because of the disparity between the power in E versus B modes in our universe, such a property would leave a signature on the EB cross-correlation spectrum, from which β can be determined (Komatsu 2022). Diego-Palazuelos et al. (2022) have used the Planck Data Release 4 data to infer a value of $\beta = 0.30° \pm 0.11°$. Future experiments will be able to test and refine this measurement, possibly placing constraints on the physics of dark matter.

7.6.5 Future Observations

The quest for measurement of a B-mode spectrum from inflationary gravitational waves continues. Ground and balloon-borne instruments such as SPIDER (Ade et al. 2022), BICEP/Keck (Tristram et al. 2022), SPT (Camphuis et al. 2025), and

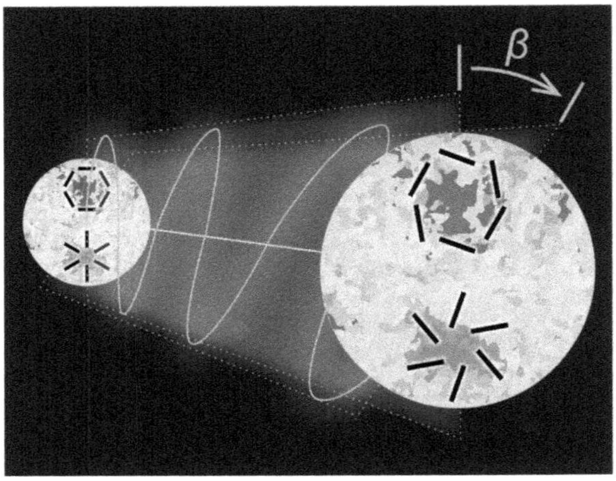

Figure 7.15. Circular cosmic birefringence that is sourced by chirality from dark energy can cause a rotation of the CMB polarization angles by an angle β.

CLASS (Eimer et al. 2024) continue to operate. The Simons Observatory (Ade et al. 2019) and new instrumentation on the South Pole Telescope (Martsen et al. 2025) will significantly increase the ground-based sensitivity to CMB polarization. Finally, LiteBIRD (Remazeilles et al. 2024) would provide a space-borne platform for all-sky measurements of the CMB polarization to unprecedented sensitivity. Future CMB polarization measurements will work in concert with information about the evolution of large scale structure from survey telescopes such as EUCLID, DESI, the Vera C. Rubin Observatory, and the Nancy Grace Roman Space Telescope to further test and refine our understanding of cosmology.

7.7 Interferometric Polarimetry

Though we have focused on single-dish telescopes, future polarimetric measurements will rely on polarimetry at multiple scales. Interferometric polarization plays a critical role at the finest angular scales, so we conclude with a quick look at some exciting results. Figure 7.16 shows multiscale polarimetry from the JCMT (Matthews et al. 2009), the Combined Array for Research in Millimeter Astronomy (CARMA) (Hull et al. 2014), the Submillimeter Telescope (SMT), and the Atacama Large Millimeter Array (ALMA; Hull et al. 2017). The subarcsecond resolution of ALMA at 870 μm enables the probe of magnetic field information at the scale of 140 AU.

The Event Horizon Telescope is a synthetic aperture interferometer that combines telescopes operating at a wavelength of λ=1.35 mm (230 GHz) (Event Horizon Telescope Collaboration et al. 2019) and achieves an angular resolution of $\sim$25 μas. It has imaged supermassive black holes in M87 and in our own Milky Way (Sgr A*) in both total intensity and polarimetry. For the latter, the polarization traces magnetic fields due to the emission of synchrotron radiation near the event horizon. Figure 7.17 shows images of the inferred magnetic field from M87 (Event Horizon

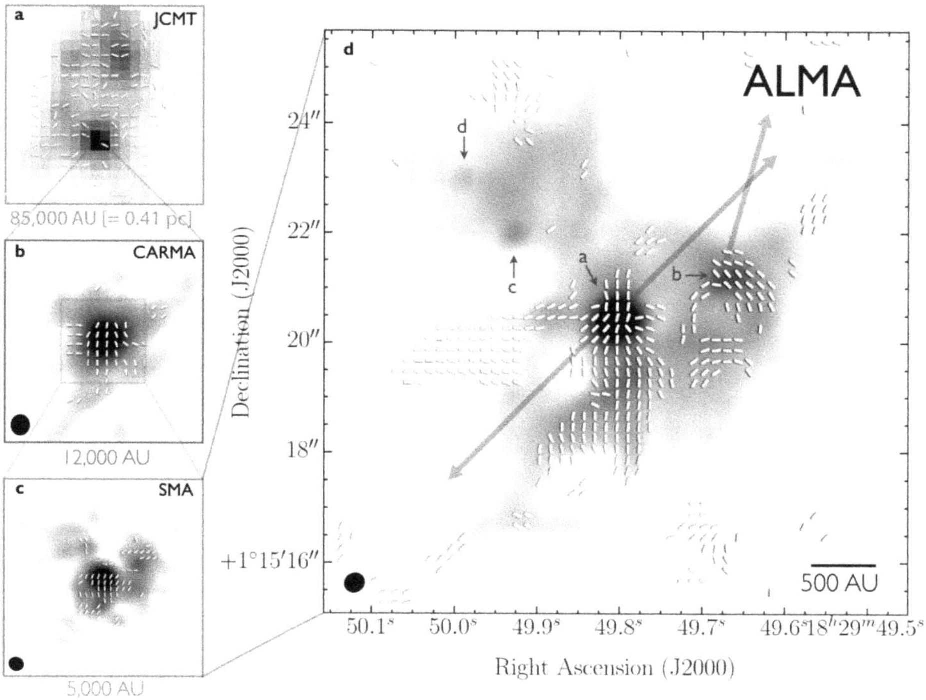

Figure 7.16. Multi-scale polarimetry progressively reveals magnetic field structure in the Class 0 Protostellar Source Serpens SMM1 (Hull et al. 2017).

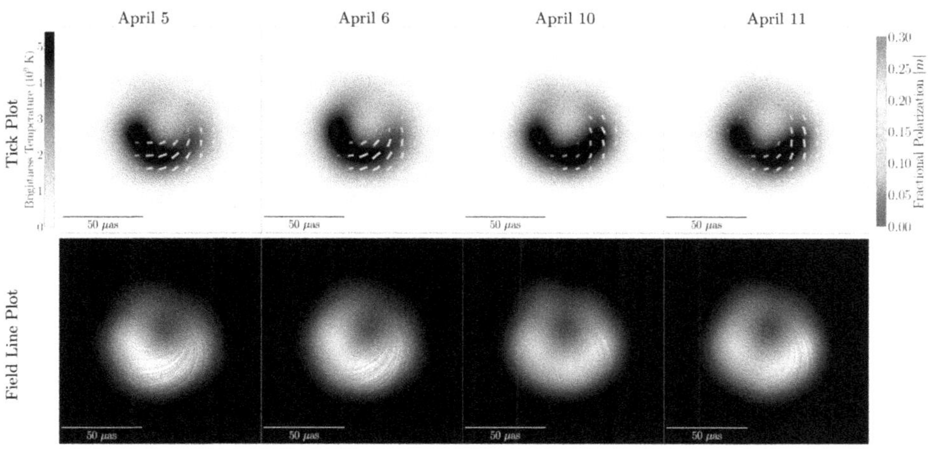

Figure 7.17. Polarization of the emission surrounding the black hole in M87 is shown (Event Horizon Telescope Collaboration et al. 2021a).

Telescope Collaboration et al. 2021a, 2021b). Continued observations will continue to refine our understanding of the dynamical importance of magnetic fields in the immediate vicinity of supermassive black holes.

References

Abazajian, K. N., Adshead, P., Ahmed, Z., et al. 2016, arXiv:1610.02743 2016arXiv161002743A
Abitbol, M. H., Ahmed, Z., Barron, D., et al. 2017, arXiv:1706.02464 2017arXiv170602464A
Adachi, S., Adkins, T., Aguilar Faúndez, M. A. O., et al. 2022, ApJ, 931, 101
Ade, P., Aguirre, J., Ahmed, Z., et al. 2019, JCAP, 2019, 056
Ade, P. A. R., Amiri, M., Benton, S. J., et al. 2022, ApJ, 927, 174
Albrecht, A., & Steinhardt, P. J. 1982, PhRvL, 48, 1220
Alexander, S., & McDonough, E. 2019, PhLB, 789, 197
Alfvén, H. 1942, Natur, 150, 405
Andersson, B. G., Karoly, J., Bastien, P., et al. 2024, ApJ, 963, 76
Andersson, B. G., Lazarian, A., & Vaillancourt, J. E. 2015, ARA&A, 53, 501
Andersson, B. G., Lopez-Rodriguez, E., Medan, I., et al. 2022, ApJ, 931, 80
Ashton, P. C., Ade, P. A. R., Angilè, F. E., et al. 2018, ApJ, 857, 10
Bally, J., Cunningham, N. J., Moeckel, N., et al. 2011, ApJ, 727, 113
Barnett, S. J. 1915, PhRv, 6, 239
Beck, R. 2015, A&ARv, 24, 4
Becklin, E. E., & Neugebauer, G. 1967, ApJ, 147, 799
Bennett, C. L., Halpern, M., Hinshaw, G., et al. 2003, ApJS, 148, 1
Bennett, C. L., Larson, D., Weiland, J. L., et al. 2013, ApJS, 208, 20
Beresnyak, A., & Lazarian, A. 2019, Turbulence in Magnetohydrodynamics (New York: De Gruyter)
Bethell, T. J., Chepurnov, A., Lazarian, A., & Kim, J. 2007, ApJ, 663, 1055
BICEP2 CollaborationAde, P. A. R., Aikin, R. W., Barkats, D., et al. 2014, PhRvL, 112, 241101
Borlaff, A. S., Lopez-Rodriguez, E., Beck, R., et al. 2023, ApJ, 952, 4
Borlaff, A. S., Lopez-Rodriguez, E., Beck, R., et al. 2021, ApJ, 921, 128
Butterfield, N. O., Guerra, J. A., Chuss, D. T., et al. 2024, ApJ, 968, 63
Camphuis, E., Quan, W., Balkenhol, L., et al. 2025, arXiv:2506.20707 2025arXiv250620707C
Chandrasekhar, S., & Fermi, E. 1953, ApJ, 118, 113
Chuss, D. T., Andersson, B. G., Bally, J., et al. 2019, ApJ, 872, 187
Chuss, D. T., Davidson, J. A., Dotson, J. L., et al. 2003, ApJ, 599, 1116
Clark, S. E., & Hensley, B. S. 2019, ApJ, 887, 136
Clemens, D. P., Leach, R. W., Barvainis, R., & Kane, B. D. 1990, PASP, 102, 1064
Cox, E. G., Novak, G., Chuss, D. T., et al. 2025, ApJ, 981, 132
Crutcher, R. M., Nutter, D. J., Ward-Thompson, D., & Kirk, J. M. 2004, ApJ, 600, 279
Cudlip, W., Furniss, I., King, K. J., & Jennings, R. E. 1982, MNRAS, 200, 1169
Davis, , Leverett, J., & Greenstein, J. L. 1951, ApJ, 114, 206
Davis, L. 1951, PhRv, 81, 890
Diego-Palazuelos, P., Eskilt, J. R., Minami, Y., et al. 2022, PhRvL, 128, 091302
Dodelson, S. 2003, Modern Cosmology (New York: Elsevier)
Dolginov, A. Z., & Mitrofanov, I. G. 1976, Ap&SS, 43, 291
Dotson, J. L. 1996, ApJ, 470, 566
Dowell, C. D. 1997, ApJ, 487, 237
Dragovan, M. 1986, ApJ, 308, 270
Draine, B. T. 2011, Physics of the Interstellar and Extragalactic Medium (Princeton, NJ: Princeton Univ. Press)
Draine, B. T., & Fraisse, A. A. 2009, ApJ, 696, 1
Draine, B. T., & Weingartner, J. C. 1997, ApJ, 480, 633
Eimer, J. R., Li, Y., Brewer, M. K., et al. 2024, ApJ, 963, 92

Event Horizon Telescope Collaboration, Akiyama, K., Alberdi, A., Alef, W., et al. 2019, ApJL, 875, L2
Event Horizon Telescope Collaboration, Akiyama, K., Algaba, J. C., Alberdi, A., et al. 2021a, ApJL, 910, L12
Event Horizon Telescope Collaboration, Akiyama, K., Algaba, J. C., Alberdi, A., et al. 2021b, ApJL, 910, L13
Faber, T. 1995, Fluid Dynamics for Physicists (Cambridge: Cambridge Univ. Press)
Fissel, L. M., Ade, P. A. R., Angilè, F. E., et al. 2010, Proc. SPIE, 7741, 77410E
Fixsen, D. J. 2009, ApJ, 707, 916
Friberg, P., Bastien, P., Berry, D., et al. 2016, Proc. SPIE, 9914, 991403
Gandilo, N. N., Ade, P. A. R., Angilè, F. E., et al. 2016, ApJ, 824, 84
Ghez, A. M., Salim, S., Weinberg, N. N., et al. 2008, ApJ, 689, 1044
Gillessen, S., Eisenhauer, F., Trippe, S., et al. 2009, ApJ, 692, 1075
Greaves, J. S., Murray, A. G., & Holland, W. S. 1994, A&A, 284, L19
Guan, Y., Clark, S. E., Hensley, B. S., et al. 2021, ApJ, 920, 6
Guerra, J. A., Chuss, D. T., Dowell, C. D., et al. 2021, ApJ, 908, 98
Guerra, J. A., Lopez-Rodriguez, E., Chuss, D. T., Butterfield, N. O., & Schmelz, J. T. 2023, AJ, 166, 37
Guillet, V., Fanciullo, L., Verstraete, L., et al. 2018, A&A, 610, A16
Guth, A. H. 1981, PhRvD, 23, 347
Hall, J. S. 1949, Sci, 109, 166
Han, J. L., Manchester, R. N., Lyne, A. G., Qiao, G. J., & van Straten, W. 2006, ApJ, 642, 868
Harper, D. A., Runyan, M. C., Dowell, C. D., et al. 2018, JAI, 7, 1840008
Hensley, B. S., & Draine, B. T. 2023, ApJ, 948, 55
Heywood, I., Rammala, I., Camilo, F., et al. 2022, ApJ, 925, 165
Hildebrand, R. H., Dotson, J. L., Dowell, C. D., Schleuning, D. A., & Vaillancourt, J. E. 1999, ApJ, 516, 834
Hildebrand, R. H., Dragovan, M., & Novak, G. 1984, ApJL, 284, L51
Hildebrand, R. H., Kirby, L., Dotson, J. L., Houde, M., & Vaillancourt, J. E. 2009, ApJ, 696, 567
Hiltner, W. A. 1949, Sci, 109, 165
Hoang, T., Tram, L. N., Lee, H., & Ahn, S.-H. 2019, NatAs, 3, 766
Holland, W. S., Duncan, W., Kelly, B. D., et al. 2003, Proc. SPIE, 4855, 1
Hollenbach, D. J., & Thronson, H. A. 1986, Summer School on Interstellar Processes: Abstracts of Contributed Papers, 88342 (Mountain View, CA: NASA Ames Research Center)
Houde, M., Dowell, C. D., Hildebrand, R. H., et al. 2004, ApJ, 604, 717
Houde, M., Hull, C. L. H., Plambeck, R. L., Vaillancourt, J. E., & Hildebrand, R. H. 2016, ApJ, 820, 38
Houde, M., Rao, R., Vaillancourt, J. E., & Hildebrand, R. H. 2011, ApJ, 733, 109
Houde, M., Vaillancourt, J. E., Hildebrand, R. H., Chitsazzadeh, S., & Kirby, L. 2009, ApJ, 706, 1504
Hu, W., & White, M. 1997, NewA, 2, 323
Hull, C. L. H., Girart, J. M., & Tychoniec, Ł. 2017, ApJ, 847, 92
Hull, C. L. H., Plambeck, R. L., Kwon, W., et al. 2014, ApJS, 213, 13
Hwang, J., Kim, J., Pattle, K., et al. 2021, ApJ, 913, 85
Inomata, K., & Kamionkowski, M. 2019, PhRvD, 99, 043501
Jackson, J. 1972, Classical Electrodynamics (2nd ed; New York: Wiley)
Jones, T. J., Dowell, C. D., Lopez Rodriguez, E., et al. 2019, ApJL, 870, L9
Jones, T. J., Kim, J.-A., Dowell, C. D., et al. 2020, AJ, 160, 167

Jow, D. L., Hill, R., Scott, D., et al. 2018, MNRAS, 474, 1018
Karoly, J., Ward-Thompson, D., Pattle, K., et al. 2025, ApJL, 982, L22
Kim, S.-H., & Martin, P. G. 1995, ApJ, 444, 293
Kogut, A., Spergel, D. N., Barnes, C., et al. 2003, ApJS, 148, 161
Komatsu, E. 2022, NatRP, 4, 452
Kounkel, M., Hartmann, L., Loinard, L., et al. 2017, ApJ, 834, 142
Kovac, J. M., Leitch, E. M., Pryke, C., et al. 2002, Natur, 420, 772
Kruijssen, J. M. D., Dale, J. E., & Longmore, S. N. 2015, MNRAS, 447, 1059
Lazarian, A., & Hoang, T. 2007a, MNRAS, 378, 910
Lazarian, A., & Hoang, T. 2007b, ApJL, 669, L77
Lazarian, A., & Yuen, K. H. 2018, ApJ, 853, 96
Lee, D., Berthoud, M., Chen, C.-Y., et al. 2021, ApJ, 918, 39
Lee, D., Chen, C.-Y., Novak, G., et al. 2024, ApJ, 972, 102
Lee, S.-S., Kang, S., Byun, D.-Y., et al. 2015, ApJL, 808, L26
Leitch, E. M., Kovac, J. M., Pryke, C., et al. 2002, Natur, 420, 763
Lewis, A., & Challinor, A. 2011, Astrophysics Source Code Library, ascl:1102.026
Li, Y., Eimer, J. R., Appel, J. W., et al. 2025, ApJ, 986, 111
Liddle, A. R., & Lyth, D. H. 2000, (Cambridge: Cambridge Univ. Press)
Lietzow-Sinjen, M., Wolf, S., & Brunngräber, R. 2025, A&A, 694, A190
Linde, A. D. 1982, PhLB, 108, 389
Lopez-Rodriguez, E., Antonucci, R., Chary, R.-R., & Kishimoto, M. 2018, ApJL, 861, L23
Lopez-Rodriguez, E., Beck, R., Clark, S. E., et al. 2021, ApJ, 923, 150
Lopez-Rodriguez, E., Clarke, M., Shenoy, S., et al. 2022a, ApJ, 936, 65
Lopez-Rodriguez, E., Dowell, C. D., Jones, T. J., et al. 2020, ApJ, 888, 66
Lopez-Rodriguez, E., Mao, S. A., Beck, R., et al. 2022b, ApJ, 936, 92
Louis, T., La Posta, A., Atkins, Z., et al. 2025, JCAP, 2025, 062
Lu, X., Liu, J., Pillai, T., et al. 2024, ApJ, 962, 39
Mainini, R., Minelli, D., Gervasi, M., et al. 2013, JCAP, 2013, 033
Mangilli, A., Aumont, J., Bernard, J. P., et al. 2019, A&A, 630, A74
Martin-Alvarez, S., Lopez-Rodriguez, E., Dacunha, T., et al. 2024, ApJ, 966, 43
Martsen, E. S., Barry, P. S., Benson, B. A., et al. 2025, ITAS, 35, TASC.2024
Matthews, B. C., McPhee, C. A., Fissel, L. M., & Curran, R. L. 2009, ApJS, 182, 143
Matthews, B. C., & Wilson, C. D. 2000, ApJ, 531, 868
Matthews, B. C., Wilson, C. D., & Fiege, J. D. 2001, ApJ, 562, 400
Matthews, T. G., Ade, P. A. R., Angilè, P. A. R., et al. 2014, ApJ, 784, 116
McKee, C. F., & Ostriker, E. C. 2007, ARAA, 45, 565
Michail, J. M., Ashton, P. C., Berthoud, M. G., et al. 2021, ApJ, 907, 46
Misawa, R., Bernard, J. P., Ade, P., et al. 2014, Proc. SPIE, 9153, 91531H
Molinari, S., Bally, J., Noriega-Crespo, A., et al. 2011, ApJL, 735, L33
Montero-Camacho, P., & Hirata, C. M. 2018, JCAP, 2018, 040
Myers, P. C., Stephens, I. W., & Coudé, S. 2024, ApJ, 962, 64
Nagy, J. M., Ade, P. A. R., Amiri, M., et al. 2017, ApJ, 844, 151
Nakano, T., & Nakamura, T. 1978, PASJ, 30, 671
Nishiyama, S., Hatano, H., Tamura, M., et al. 2010, ApJL, 722, L23
Novak, G., Chuss, D. T., Renbarger, T., et al. 2003, ApJL, 583, L83
Novak, G., Dotson, J. L., Dowell, C. D., et al. 2000, ApJ, 529, 241
Packham, C., Hough, J. H., Young, S., et al. 1996, MNRAS, 278, 406

Padilla, I. L., Eimer, J. R., Li, Y., et al. 2020, ApJ, 889, 105
Page, L., Hinshaw, G., Komatsu, E., et al. 2007, ApJS, 170, 335
Paoletti, D., Finelli, F., Valiviita, J., & Hazumi, M. 2022, PhRvD, 106, 083528
Paré, D., Butterfield, N. O., Chuss, D. T., et al. 2024, ApJ, 969, 150
Paré, D. M., Chuss, D. T., Karpovich, K., et al. 2025, ApJ, 978, 28
Pattle, K., Lai, S.-P., Di Francesco, J., et al. 2021, ApJ, 907, 88
Pattle, K., Ward-Thompson, D., Berry, D., et al. 2017, ApJ, 846, 122
Pattle, K., Ward-Thompson, D., Hasegawa, T., et al. 2018, ApJL, 860, L6
Peiris, H. V., Komatsu, E., Verde, L., et al. 2003, ApJS, 148, 213
Perry, J. P., Pattle, K., Johnstone, D., et al. 2025, MNRAS, 536, 1736
Petroff, M. A., Eimer, J. R., Harrington, K., et al. 2020, ApJ, 889, 120
Pillai, T. G. S., Clemens, D. P., Reissl, S., et al. 2020, NatAs, 4, 1195
Planck CollaborationAdam, R., Ade, P. A. R., Aghanim, N., et al. 2016a, A&A, 586, A133
Planck CollaborationAde, P. A. R., Aghanim, N., Alina, D., et al. 2015, A&A, 576, A104
Planck CollaborationAde, P. A. R., Aghanim, N., Alves, M. I. R., et al. 2016b, A&A, 586, A138
Planck CollaborationAde, P. A. R., Aghanim, N., Arnaud, M., et al. 2016c, A&A, 586, A141
Planck CollaborationAghanim, N., Akrami, Y., Alves, M. I. R., et al. 2020a, A&A, 641, A12
Planck CollaborationAghanim, N., Akrami, Y., Arroja, F., et al. 2020b, A&A, 641, A1
Planck CollaborationAghanim, N., Akrami, Y., Ashdown, M., et al. 2020c, A&A, 641, A6
Planck CollaborationAghanim, N., Akrami, Y., Ashdown, M., et al. 2020d, A&A, 641, A5
Remazeilles, M., Douspis, M., Rubiño-Martín, J., et al. 2024, JCAP, 2024, 026
Ritacco, A., Ponthieu, N., Catalano, A., et al. 2017, A&A, 599, A34
Rybicki, G. B., & Lightman, A. P. 1979, Radiative Processes in Astrophysics (New York: Wiley)
Santos, F. P., Chuss, D. T., Dowell, C. D., et al. 2019, ApJ, 882, 113
Schleuning, D. A. 1998, ApJ, 493, 811
Seljak, U., & Zaldarriaga, M. 1996, ApJ, 469, 437
Skalidis, R., & Tassis, K. 2021, A&A, 647, A186
Smoot, G. F., Bennett, C. L., Kogut, A., et al. 1992, ApJL, 396, L1
Soler, J. D., Hennebelle, P., Martin, P. G., et al. 2013, ApJ, 774, 128
Spruit, H. C. 2013, arXiv:1301.5572 2013arXiv1301.5572S
Tram, L. N., Bonne, L., Hu, Y., et al. 2023, ApJ, 946, 8
Tram, L. N., Hoang, T., Lee, H., et al. 2021a, ApJ, 906, 115
Tram, L. N., Lee, H., Hoang, T., et al. 2021b, ApJ, 908, 159
Tristram, M., Banday, A. J., Górski, K. M., et al. 2022, PhRvD, 105, 083524
Vaillancourt, J. E. 2002, ApJS, 142, 53
Ward-Thompson, D., Pattle, K., Bastien, P., et al. 2017, ApJ, 842, 66
Weingartner, J. C., & Draine, B. T. 2003, ApJ, 589, 289
Wu, J., Qiu, K., Poidevin, F., et al. 2024, ApJL, 977, L31
Yusef-Zadeh, F., & Morris, M. 1987a, AJ, 94, 1178
Yusef-Zadeh, F., & Morris, M. 1987b, ApJ, 320, 545
Yusef-Zadeh, F., Morris, M., & Chance, D. 1984, Natur, 310, 557
Zaldarriaga, M., & Seljak, U. 1998, PhRvD, 58, 023003
Zeng, L., Bennett, C. L., Chapman, N. L., et al. 2013, ApJ, 773, 29
Zielinski, N., & Wolf, S. 2022, A&A, 659, A22

Appendix A

Maxwell's Equations

We include Maxwell's equations in differential form for convenience. Astronomy typically utilizes Gaussian units for notational simplicity in theoretical calculations; whereas, in instrumentation, SI units are generally preferred from measurement and material properties book-keeping perspectives. Each strategy has its advantages and rationale, so we include both here. For a more complete treatment of Maxwell's equations and their units, the reader is directed to electrodynamics texts such as Jackson (1972).

In SI units the speed of light is related to the permittivity and permeability of the freespace by $c = 1/\sqrt{\varepsilon\mu}$, and has the standard MKS units (m s^{-1}). The SI units of permeability and permittivity have physical units of H m^{-1} and F m^{-1}, respectively. SI units are "rationalized," meaning that a factor 4π appears in the field equations (i.e., not in Maxwell's equations):

$$\nabla \cdot \mathbf{D} = \rho \qquad \nabla \times \mathbf{E} + \frac{\partial \mathbf{B}}{\partial t} = 0$$

$$\nabla \cdot \mathbf{B} = 0 \qquad \nabla \times \mathbf{H} - \frac{\partial \mathbf{D}}{\partial t} = \mathbf{J}$$

In Gaussian units, the permeability and permittivity of free space are each unity, and as a result factors of c appear in Maxwell's equations directly. Consequently, **E** and **D** have the same units; as do **B** and **H**. In this setting, the material properties are carried by the *relative permittivity and permeability* of the media. Gaussian units are "unrationalized," meaning that a factor 4π appears in Maxwell's equations (i.e., not in field equations):

$$\nabla \cdot \mathbf{D} = 4\pi\rho \qquad \nabla \times \mathbf{E} + \frac{1}{c}\frac{\partial \mathbf{B}}{\partial t} = 0$$

$$\nabla \cdot \mathbf{B} = 0 \qquad \nabla \times \mathbf{H} - \frac{1}{c}\frac{\partial \mathbf{D}}{\partial t} = \frac{4\pi}{c}\mathbf{J}$$

doi:10.1088/2514-3433/ae1022ch8

Reference

Jackson, J. 1972, Classical Electrodynamics (2nd ed.; New York: Wiley)

Astrophysical Polarimetry
Far-infrared through millimeter wavelengths
David T Chuss and Edward J Wollack

Appendix B

The Curl of a Vector Product

Because it is relevant to the magnetohydrodynamic (MHD) Navier–Stokes equation, for the convenience of the reader, we provide a proof of

$$\nabla \times (\mathbf{A} \times \mathbf{B}) = \mathbf{A}(\nabla \cdot \mathbf{B}) - \mathbf{B}(\nabla \cdot \mathbf{A}) + (\mathbf{B} \cdot \nabla)\mathbf{A} - (\mathbf{A} \cdot \nabla)\mathbf{B}.$$

To do this, we utilize summation notation and write out this expression in terms of the Levi–Civita symbol,

$$\nabla \times (\mathbf{A} \times \mathbf{B}) = \sum_{ijk} \varepsilon_{ijk} \hat{\mathbf{e}}_i \frac{\partial}{\partial x_j} (\mathbf{A} \times \mathbf{B})_k = \sum_{ijk} \varepsilon_{ijk} \hat{\mathbf{e}}_i \frac{\partial}{\partial x_j} \left(\sum_{lmn} \varepsilon_{lmn} \hat{\mathbf{e}}_l A_m B_n \right)_k .$$

We can take the kth term of the final factor by only including the kth term of the summation,

$$= \sum_{ijk} \varepsilon_{ijk} \hat{\mathbf{e}}_i \frac{\partial}{\partial x_j} \left(\sum_{lmn} \varepsilon_{kmn} A_m B_n \right).$$

Next, we re-order the summed indices,

$$= \sum_i \hat{\mathbf{e}}_i \sum_j \frac{\partial}{\partial x_j} \left(\sum_{mn} A_m B_n \right) \sum_k \varepsilon_{ijk} \varepsilon_{klm}.$$

Next, we can use the fact that

$$\sum_k \varepsilon_{ijk} \varepsilon_{kmn} = \delta_{im} \delta_{jn} - \delta_{in} \delta_{jm} \tag{B.1}$$

to simplify the expression,

doi:10.1088/2514-3433/ae1022ch9

$$= \sum_i \hat{\mathbf{e}}_i \sum_j \frac{\partial}{\partial x_j}\left(\sum_{mn} A_m B_n\right)(\delta_{im}\delta_{jn} - \delta_{in}\delta_{jm})$$

$$= \sum_i \hat{\mathbf{e}}_i \left[\sum_j \frac{\partial}{\partial x_j}\left(\sum_{mn} A_m B_n\right)\delta_{im}\delta_{jn} - \sum_j \frac{\partial}{\partial x_j}\left(\sum_{mn} A_m B_n\right)\delta_{in}\delta_{jm}\right]$$

$$= \sum_i \hat{\mathbf{e}}_i \left[\sum_j \frac{\partial}{\partial x_j}(A_i B_j) - \sum_j \frac{\partial}{\partial x_j}(A_j B_i)\right]$$

$$= \sum_i \hat{\mathbf{e}}_i \left[A_i \sum_j \frac{\partial B_j}{\partial x_j} + \sum_j B_j \frac{\partial A_i}{\partial x_j} - \sum_j A_j \frac{\partial B_i}{\partial x_j} - B_i \sum_j \frac{\partial A_j}{\partial x_j}\right]$$

$$= \sum_i \hat{\mathbf{e}}_i A_i \sum_j \frac{\partial B_j}{\partial x_j} - \sum_i \hat{\mathbf{e}}_i B_i \sum_j \frac{\partial A_j}{\partial x_j} + \sum_i \hat{\mathbf{e}}_i \sum_j \left(B_j \frac{\partial A_i}{\partial x_j} - A_j \frac{\partial B_i}{\partial x_j}\right).$$

Putting things back in vector notation, we find

$$\mathbf{A}(\nabla \cdot \mathbf{B}) - \mathbf{B}(\nabla \cdot \mathbf{A}) + (\mathbf{B} \cdot \nabla)\mathbf{A} - (\mathbf{A} \cdot \nabla)\mathbf{B}.$$

In the limit of high conductivity, this identity applies to MHD equations explored in Chapter 7 through the expression

$$\nabla \times (\mathbf{v} \times \mathbf{B}) = \mathbf{v}(\nabla \cdot \mathbf{B}) - \mathbf{B}(\nabla \cdot \mathbf{v}) + (\mathbf{B} \cdot \nabla)\mathbf{v} - (\mathbf{v} \cdot \nabla)\mathbf{B},$$

where $\mathbf{v}$ is the velocity and now $\mathbf{B}$ is the magnetic field. From Maxwell's equations, we know that $\nabla \cdot \mathbf{B} = 0$. The velocity is divergence free ($\nabla \cdot \mathbf{v} = 0$) if the fluid is incompressible, which is a common approximation in molecular clouds and the diffuse ISM. From this, we arrive at the result,

$$\nabla \times (\mathbf{v} \times \mathbf{B}) = (\mathbf{B} \cdot \nabla)\mathbf{v} - (\mathbf{v} \cdot \nabla)\mathbf{B}.$$

Appendix C

Polar and Axial Vectors in Electrodynamics

In electromagnetic theory, symmetry operations in the spatial and temporal coordinates have different effects on the physical field quantities. Reflection (mirror), inversion (parity), and time-reversal are mathematical transformations that physically influence the field solutions in distinctly different ways, primarily because the electric and magnetic field solutions differ fundamentally from a symmetry perspective as required by Maxwell's equations. The electric field vectors, $\mathbf{E}(\mathbf{r}, t)$, are polar (true) vectors[1]. The magnetic field solutions, $\mathbf{B}(\mathbf{r}, t) = \boldsymbol{\nabla} \times \mathbf{A}$, are axial vectors (pseudovectors)[2]. Expression of the magnetic field as the curl of the magnetic vector potential, $\mathbf{A}(\mathbf{r}, t)$, is a direct consequence of $\boldsymbol{\nabla} \cdot \mathbf{B} = 0$. These symmetry transformations are illustrated in Figure C.1 and defined as follows:

- **Reflection** (or mirror) symmetry describes invariance encountered when one half of a system is the mirror image of the other across a specific spatial plane (e.g., left–right or bilateral symmetry of a butterfly or similar). The relative orientation of the parallel and perpendicular components of electromagnetic fields to the symmetry plane determine the outcome of the transformation on the system response. For a planar interface the Fresnel equations directly encode the boundary conditions and enable computation of the electromagnetic response. Image theory can have utility for highly-symmetric structures by allowing the replacement of sources and boundaries with equivalent sources and virtual images to create an identical field distribution with a known (or calculable) solution. Similarly, electric and magnetic mirror symmetries (or "perfect" boundaries) enable a reduction of resources required in numerical electromagnetic simulations.

[1] Polar vectors are anti-symmetrical (symmetrical) with respect to a reflection through a perpendicular (parallel) plane. Polar or "true" vectors are simply vectors.

[2] Axial vectors are anti-symmetrical (symmetrical) with respect to a reflection through a parallel (perpendicular) plane. Axial or "pseudo" vectors are formally anti-symmetric tensors of rank two.

doi:10.1088/2514-3433/ae1022ch10

- **Inversion** (or parity) is a global symmetry transformation that inverts all spatial coordinates through a point (e.g., flipping all three spatial coordinates, $\mathbf{r} \rightarrow -\mathbf{r}$, through the origin). In electromagnetics, the electric fields reverse direction under inversion, $\mathbf{E}(\mathbf{r}, t) = -\mathbf{E}(-\mathbf{r}, t)$, but the magnetic fields do not, $\mathbf{B}(\mathbf{r}, t) = \mathbf{B}(-\mathbf{r}, t)$. The laws of classical electromagnetism are invariant under parity transformation[3].

- **Time-reversal** is a global symmetry transformation that reverses the direction of time coordinate (i.e., $t \rightarrow -t$), which can be interpreted as arising from changing the direction of charge motion in the system (i.e., reversing the velocity vector of charge carriers, $\mathbf{v} \rightarrow -\mathbf{v}$, or the current density vector, $\mathbf{J} \rightarrow -\mathbf{J}$), which is the product of the charge number density and velocity. In electromagnetics, the electric fields do not reverse direction under time-reversal, $\mathbf{E}(\mathbf{r}, t) = \mathbf{E}(\mathbf{r}, -t)$, but the magnetic fields do, $\mathbf{B}(\mathbf{r}, t) = -\mathbf{B}(\mathbf{r}, -t)$[4]. In "reciprocal" electromagnetic media (i.e., a time-invariant, linear, non-magnetic systems), the fields at the source and observer locations are invariant under interchange due to time-reversal symmetry. In antenna and circuit theory, this concept is widely applied to simplify calculations by swapping the roles of the transmitting and receiving elements to enable computation (and/or measurement) of the desired electromagnetic response.

In addition to the symmetry operations discussed here, other examples include rotational, periodicity, and self-complementarity. To learn more about practical applications of electromagnetic symmetries and the design possibilities enabled, several references are highly recommended. Chen (1983) follows an interesting approach, which primarily concentrates on the use of coordinate-free notation rather than applications to elucidate Maxwell's equations. This relatively compact notational approach provides a path to understand and appreciate the functional form of the relations found in electromagnetics. The implications of these symmetries are explored in waveguide structures for discrete modes by Montgomery et al. (1948) and for adiabatic propagation by Collin (1992). The monograph on broad band antennas by Rumsey (1966) explores the symmetry conditions leading to frequency independent structures and self-complementary. More advanced insights regarding symmetry and complementarity in planar structures can be found in

[3] Polar and axial vectors transform identically under rotation. However, for the combined operations of reflection-inversion an axial vector encounters an additional sign change relative to a polar vector. This arises because a pseudovector has a handedness (chirality), which is reversed under this sequence of operations. While not necessarily profound, this underlies the fundamental difference encountered between the fields encountered in classical electromagnetics.

[4] Generally speaking electromagnetism is unchanged under simultaneous *charge* (C) conjugation (i.e., flipping the sign of charges from positive to negative), *parity* (P), and *time* (T) reversal. This combination of symmetry operators is referred to as CPT symmetry. It is observed to be an exact symmetry in nature as anticipated from Lorentz invariance.

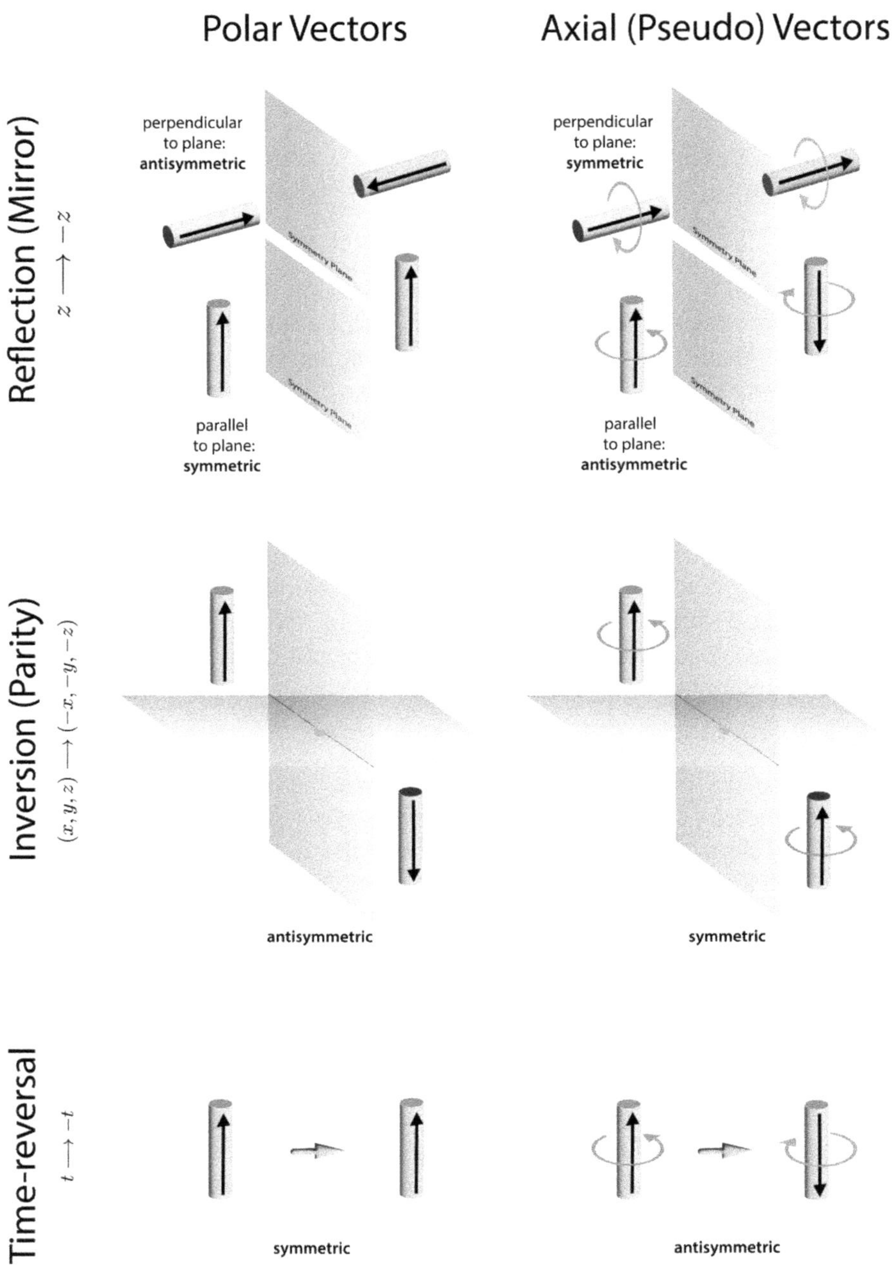

Figure C.1. Reflection, inversion, and time-reversal symmetries are illustrated in terms of their effect on polar and axial vectors. In the top row, the symmetry plane for reflection is labeled, and the effect of the transformation on both entities is illustrated for orientations both parallel and perpendicular to the symmetry axis. In the middle row, the inversion point is represented by the small orange sphere. The small right arrow in the third row represents time-reversal.

Deschamps (1959); Booker (1946); Mackay (1989). This approach also provides a path to understand and appreciate the functional form of the relations found in electromagnetics. For discussion of electromagnetic duality symmetry and its relationship to circular polarization and optical helicity conservation, see Calkin (1965) and Cameron (2012).

References

Booker, H. G. 1946, J. Inst. Electr. Eng., Part 3A, 93, 620

Calkin, M. G. 1965, American Journal of Physics, 33, 958

Cameron, R. P., & Barnett, S. M. 2012, New Journal of Physics, 14, 123019

Chen, H. C. 1983, Theory of Electromagnetic Waves: A Coordinate-Free Approach (Electrical Engineering Series) (McGraw-Hill)

Collin, R. E. 1992, Foundations for Microwave Engineering 2nd ed. (McGraw-Hill)

Deschamps, G. 1959, IRE Trans. Antennas Propag., 7, 371

Mackay, A. 1989, ElL, 25, 1624

Montgomery, C. G., Dicke, R. H., & Purcell, E. M. 1948, Principles of Microwave Circuits (MIT Radiation Laboratory Series) Vol 8, (McGraw-Hill)

Rumsey, V. 1966, Frequency Independent Antennas (Electrical Science Series) (Academic Press)

www.ingramcontent.com/pod-product-compliance
Ingram Content Group UK Ltd.
Pitfield, Milton Keynes, MK11 3LW, UK
UKHW051924100926
8556IPUK00021B/1354

9 780750 327541